Unbuilt Environments

The Nature | History | Society series is devoted to the publication of high-quality scholarship in environmental history and allied fields. Its broad compass is signalled by its title: *nature* because it takes the natural world seriously; *history* because it aims to foster work that has temporal depth; and *society* because its essential concern is with the interface between nature and society, broadly conceived. The series is avowedly interdisciplinary and is open to the work of anthropologists, ecologists, historians, geographers, literary scholars, political scientists, sociologists, and others whose interests resonate with its mandate. It offers a timely outlet for lively, innovative, and well-written work on the interaction of people and nature through time in North America.

General Editor: Graeme Wynn, University of British Columbia

A list of titles in the series appears at the end of the book.

Unbuilt Environments

Tracing Postwar Development in Northwest British Columbia

JONATHAN PEYTON

FOREWORD BY GRAEME WYNN

UBC Press • Vancouver • Toronto

25 24 23 22 21 20 19 18 17 5 4 3 2 1

Printed in Canada on FSC-certified ancient-forest-free paper (100% post-consumer recycled) that is processed chlorine- and acid-free.

Library and Archives Canada Cataloguing in Publication

Peyton, Jonathan, author
Unbuilt environments : tracing postwar development in northwest British Columbia / Jonathan Peyton.

(Nature, history, society)
Includes bibliographical references and index.
Issued in print and electronic formats.
ISBN 978-0-7748-3304-2 (hardback). – ISBN 978-0-7748-3305-9 (pbk.)
ISBN 978-0-7748-3306-6 (pdf). – ISBN 978-0-7748-3307-3 (epub)
ISBN 978-0-7748-3308-0 (mobi)

1. Natural resources – Environmental aspects – British Columbia – Case studies. 2. Natural resources – Social aspects – British Columbia – Case studies. 3. Infrastructure (Economics) – Environmental aspects – British Columbia – Case studies. 4. Infrastructure (Economics) – Social aspects – British Columbia – Case studies. 5. Economic development – Environmental aspects – British Columbia – Case studies. 6. Economic development – Social aspects – British Columbia – Case studies. 7. British Columbia – Environmental conditions – Case studies. 8. British Columbia – Social conditions – Case studies. I. Title. II. Series: Nature, history, society

HC117.B8P49 2017 330.9711'8504 C2016-905619-8
C2016-905620-1

Canadä

UBC Press gratefully acknowledges the financial support for our publishing program of the Government of Canada (through the Canada Book Fund), the Canada Council for the Arts, and the British Columbia Arts Council.

This book has been published with the help of a grant from the Canadian Federation for the Humanities and Social Sciences, through the Awards to Scholarly Publications Program, using funds provided by the Social Sciences and Humanities Research Council of Canada, and with the help of the University of British Columbia through the K.D. Srivastava Fund.

UBC Press
The University of British Columbia
2029 West Mall
Vancouver, BC V6T 1Z2
www.ubcpress.ca

Contents

Tables and Figures

Tables

Figures

FOREWORD

How Shall We Live?

Graeme Wynn

The Stikine is said to be one of the last truly wild rivers of British Columbia, although it reaches the sea in southern Alaska, midway between Juneau and the Canadian port of Prince Rupert. The river rises in a subalpine basin, high in the Cassiar Mountains, from which the Skeena and the Nass Rivers also flow. In the language of the Indigenous Tahltan whose territory this is, the meadows that give rise to the rivers are known as *Klabona* (or headwaters). In recent years the Klabona Keepers Elders Society and others have deified this "highly traditional use area" as the Sacred Headwaters. Immediately north and east is the area known to mapmakers as the Spatsizi Plateau Wilderness, bound on its inland edge by the upper reaches of the Stikine River; some have described this as "Canada's Serengeti," to mark – by powerful allusion to the game-rich plains of equatorial east-Africa – the significant populations of Stone sheep, grizzly bear, caribou, and wolf found there. Flowing, for much of its 600-kilometre course, through mountainous terrain, the Stikine has carved deep gorges through which it runs swift and turbulent beneath snow- and ice-capped peaks to its braided estuary. When the famed American naturalist and wilderness-lover John Muir travelled the lower third of the river in 1879, he counted some three hundred glaciers along his way, and likened the scenes through which he passed to "a Yosemite that was a hundred miles long."[1]

Wade Davis – ethnographer, photographer and filmmaker, sometime explorer-in-residence at the National Geographic Society in Washington,

DC, prolific writer, stirring lecturer, world traveller, and, according to David Suzuki, "a rare combination of scientist, scholar, poet and passionate defender of all of life's diversity" – also knows and loves the Stikine. He came to this corner of northern British Columbia to take a job as a provincial park ranger in the newly established Spatsizi wilderness area and, emotionally at least, he never left. In two four-month summers in the late 1970s, he saw a dozen people. This, he realized, was about as remote a place as any on earth. In the 1970s, as now, a narrow ribbon of gravel and tarmac running along the flanks of the Coast Mountains to Yukon Territory was the only maintained road in an area as large as the state of Oregon and almost twice the size of the United Kingdom. Before transcontinental railroads spanned North America, and the Panama Canal eliminated the need to navigate the Straits of Magellan, this area was as far from Britain as any part of that enormous Empire on which the sun never set. Its landscapes – including mountains rising above 2,500 metres, subalpine meadows, pristine lakes, and the Grand Canyon of the Stikine, seventy-two kilometres long and more than three hundred metres deep – are often hailed as among the most spectacular and beautiful in the world.[2]

Over the years, Davis has grown increasingly fond of this place and his neighbours. Here, in 1987, he made his summer home on Ealue Lake. He looks on the surrounding area, characterized by one reporter as "Canada's own Eden," as the most "stunningly wild place he has ever been."[3] Becoming acquainted with it meant listening to the stories of local people (such as his mentor Alex Jack, who spoke a handful of languages reflecting the mix of people Indigenous to the area, as well as English learned from a newcomer priest). These were people raised on the land, illiterate but wise, whose "intuitions and sensibilities about the natural world … [Davis] could only admire." In time, he came to appreciate that the entire Iskut Valley, into which Ealue Lake drains, "is a garden … brought alive and given meaning by generation upon generation of indigenous men and women, living, breathing, and dying, leaving in their wake history, culture and myth."[4] Like many of us, Davis has explored and photographed his treasured "backyard" in many lights and various moods. And if few of us can think of such a space as he enjoys ("where humans are scarce and jaw-dropping scenery is plentiful") as home, we might nonetheless share his sense that our patch of familiar earth is something to cherish and protect.[5] Indeed, the fences, hedges, walls, gates, and even occasional "No Trespassing" signs that abound in suburb after suburb, on the further fringes of every city in North America and across the countryside beyond, are material evidence of our strong desire to do just that.

There are few boundary markers or "Keep Out" signs to speak of in the Stikine, however. Isolation has been the saving grace of this country – at least until recently. No European vessel reached the mouth of the Stikine River until the very end of the eighteenth century; Robert Campbell of the Hudson's Bay Company crossed the Rockies and established a post at Dease Lake in 1837, but this outpost lasted barely two years. Workers intent on constructing an overland telegraph through Siberia to Europe, as well as seekers after gold, came to (and left) the region in the second half of the century, and the diseases Europeans introduced among Native populations wherever they went in the Americas took their usual devastating toll of Indigenous lives in this corner of the continent. Influenza and measles were also more disruptive of the area's traditional societies than were newcomer humans through the first half of the twentieth century. Then, shortly after the war, public-school-educated Englishman turned guide-outfitter Tommy Walker lit out from Bella Coola to establish a lodge on the Spatsizi Plateau. By his account, he sought a "sanctuary from progress" where "the unwritten laws of the frontier and ... a Christian code of ethics" prevailed. By local accounts, reported by Wade Davis, self-interest trumped Christian charity in Walker's dealings with the Metsantan people, whose hunting he feared might spoil his business. At his urging, most of them were removed from the plateau by the Indian Agent.[6]

At much the same time, the Stikine, Skeena, and Nass watersheds were being encircled by development. To the north, one hundred kilometres from the Yukon border, an offshoot of Conwest Exploration Company opened an asbestos mine at Cassiar early in the 1950s, linking it by road to the Alaska Highway at Watson Lake. A few years later, BC premier W.A.C. Bennett reached an agreement (that ultimately came to very little) with Swedish industrialist Axel Wenner-Gren to develop a vast tract of territory to the east. Dams, mines, pulpmills, and a high-speed monorail were on the agenda for this ten-million-hectare "area of awesome beauty ... [and] brilliantly coloured lakes, set in primeval forests of poplar and pine" in the Rocky Mountain Trench, which (*Life Magazine* alleged) had become known locally as Wenner-Grenland.[7] To the south, government policies, the expansion of truck-logging, and market demand quickened industry's assault on the forests and fostered the growth of Prince George as a saw- and pulp-milling centre.[8] Prospectors, some employed by US mining giant Kennecott Minerals, were also at work, staking claims through the northwest of the province. Interest in the lower reaches of the Stikine and Iskut Rivers, an extensively mineralized belt that the mining industry has called the Golden Triangle, was high. Indeed, Edward Hoagland, an American adventurer who visited the region twice in the late 1960s, found "a frank new air of rapine" there on his return in 1968.[9]

Early in the 1970s, the British Columbia government built a road through the heart of the region, from Stewart to Cassiar, linking southern parts of the province to the Alaska Highway. Development quickened. Although the New Democratic Party established the Spatsizi Plateau Wilderness Park in 1975 to protect wildlife ecosystems in the area, the protected area was smaller than proponents had hoped for, and remained vulnerable to exploitation. In the 1970s and 1980s, various groups framed grand designs for mega-development projects in the heart of the Stikine: the extension of the railway from Fort St. James to Dease Lake; a series of massive hydroelectric dams on the Iskut and Stikine Rivers; and a plan to quadruple the extent of logging in the region's southern reaches. Mining also expanded. Late in the 1980s, promoters claimed that there were six million ounces of gold to be had in the Golden Triangle. The Golden Bear, Eskay Creek, Snip, Premier, and other mines were brought into production. By the end of 2002, by the company's account, Eskay Creek had produced 2.86 million ounces of gold and 129.6 million ounces of silver.[10] Even as that mine closed in 2008, having returned $135 million to the Tahltan First Nation and immeasurably more to its shareholders, self-interest and the euphoria of a worldwide boom in commodity prices led the Mining Association of BC to claim that construction of a new transmission line to bring electricity into the region would generate $15 billion in new capital investment and produce ten thousand full-time jobs. A year later, federal and provincial governments and BC Hydro announced plans to electrify the north by building such a line from Meziadin Junction, near Stewart, to Bob Quinn Lake, south of Dease Lake. This new Northwest Transmission Line, said to run within "a few miles of more than a dozen of the world's richest untapped mineral deposits," seemed set, in the words of an Eskay Mining Corporation statement, to turn the region into "THE big story in the mining world."[11] Several years later, in 2016, this great tale remains untold. The Northwest Transmission Line was completed late in 2014, at a cost almost double the original estimates, but new mines have yet to yield their promised returns.

At one level, this is the story of the Stikine, a story of improving connections with the world beyond, of rising interest in local resources, of differing perceptions of and ambitions for the area, and of Indigenous people marginalized. In very general terms, this is a broadly familiar tale in the long history of human territorial expansion. New worlds have always been magnets for development dreams, and as the great frontier opened up through half a millennium, new islands, continents, regions, basins, and watersheds attracted adventurers, settlers, promoters, and speculators. Most of these were hard-headed, practical, and resilient people; few would readily have called themselves dreamers. But almost all of them envisaged a world different from that in which they lived. Some aspired to build better societies: "a city on a hill" watched by the world (as John Winthrop framed it in preaching to his fellow Puritans as the *Arbella* sailed towards New England in 1630), or "communities of true inspiration" (of the sort that those entering the Amana Colonies of Iowa believed they were creating in the second half of the nineteenth century). On frontiers where so much was new and almost everything remained to be determined, boosters driven by a mix of self-interest and communal ambition promoted towns, cities, colonies, and regions, burnishing public perceptions of their prospects in hopes of attracting attention, people, and investment. In the nineteenth century, progress leagues, population clubs, beautifying associations, and publicists of many stripes dreamed big and had few qualms about myth-mongering and overstatement as they vied with one another to turn hopes and visions into reality. Few, indeed, were as circumspect as the California real estate firm that claimed, in 1887, that San Diego had a population of 150,000, "only they are not all here yet."[12] In more recent times, economic development agencies, advertising companies, and property developers have tracked the same vein. Yet others – most of the men who moiled for gold around the Pacific Rim in the nineteenth century, perhaps – dreamed, almost exclusively, of finding their El Dorado, "making it rich," and retreating to more salubrious environs.

In the grand scheme of things, reveries turned into nightmares and delusions as often as they became realities. Winthrop's shining "city on a hill" was soon riven by "new and dangerous ideas"; many utopian communities, perhaps especially those that believed in celibacy, failed to span generations; the booster's tendency to describe the future as though it had arrived always stood to be confounded – as it was in an 1889 bird's-eye view of Denver, Colorado, that oriented the imposing, domed Capitol building east-west, rather than north-south as it was aligned when construction began in 1890.[13] Many a prospector was left broken in health, spirit, or ambition before he struck pay dirt. Even those who prospered sometimes found their idyll haunted by ghosts. As the Australian scientist and conservationist Tim Flannery has reminded us, future seekers were also future eaters.[14] In changing places to achieve their purposes, these individuals destabilized the lives and foreclosed the prospects of others, both human and non-human.

The devil is in these details. If northwestern British Columbia can be understood, at one level, as a remote and late frontier of European development, it is also *sui generis.* Beneath the broad narrative arc of economic and imperial expansion, there are other tales to tell, tales inflected by particular settings, specific temporal conjunctures, and the complex interplay of local actors, with provincial, national, and global influences. As one burrows down, complications arise and stories proliferate. Deciding which of these stories to tell, and how to tell them, is a challenge for those who would write a close-grained account of any circumscribed locale or more-or-less discrete society. Rising to the challenge through the pages that follow, Jonathan Peyton provides an innovative, informative rumination on recent resource and infrastructure developments in northwestern BC – and raises thought-provoking questions about the wider, cumulative significance of the "imaginative possibilities" or transformative dreams with which he engages.

Novelist Wayne Johnston has characterized his native Newfoundland as a colony of unrequited dreams, a location in which, critic Michael Collins has observed, the "shifty game of what-if and maybe and never-was" has been played out over and over again.[15] Peyton, born and bred in British Columbia, places such what-ifs, maybes, and never-weres at the very heart of his book. His interest is in the "unbuilt environments" of the Stikine. His achievement is in drawing attention not only to the ephemerality of many development projects but in pondering their implications or legacies. Still, it is well to recognize, from the outset, that Peyton adopts a somewhat broad framing of both the region and the interpretive ("unbuilt") theme that define his account.

Peyton's Stikine is both less and more than the watershed of that river. He has very little to say of its lower, Alaskan, reaches. He gives only passing consideration to the intricate tapestry of economy and society in the watershed since 1950. He does not dwell on the establishment and management of parks and special management zones there – in 2000, the BC government added 1.1 million acres of parkland to the 2.4 million already in parks between the headwaters and the Alaska border, and designated another 4.1 million acres as Special Management Zones off limits to logging, thus placing 60 percent of the Stikine-Iskut watershed under some form of conservation management.[16] Nor is Peyton's scope confined by the boundaries of the basin. Cassiar, the asbestos mining town that is the focus of his first chapter, lies beyond the northern edge of the Stikine drainage. Plans for the export of liquified natural gas from northwestern BC, discussed in Chapter 4, had global reach but bypassed the Stikine entirely. The liquefaction plant that was the hinge of this project was to have been located at Grassy Point at the mouth of the Nass River, south of the Stikine; it would have drawn gas by pipeline from Alberta and shipped it to Japan. Here, "the Stikine" is a broad, spatial signifier rather than a precise categorization of place. Peyton founds his analysis in part on recent work in political ecology that has destabilized bounded notions of place even as it attends to "specific geographies of difference" (p. 13). In this view, places "are unbounded, fragmentary, and fluid" (p. 12). They are rendered malleable by the weight of history, politics, and environmental dislocation, and made and remade through a continuing process of contestation and negotiation; as a consequence, the traditional metrics of geographical inquiry – scale, nature, landscape, environment, place – lose their rigidity.

The notion of the "unbuilt environment" also warrants further contemplation. As Peyton acknowledges, he derives his use of the term from Kathryn Oberdeck, who argues the importance of using architectural plans for never-erected buildings in the town of Kohler, Wisconsin, to "[unfold] what might have been."[17] Though she may have coined the phrase "archives of the unbuilt environment" to refer to these documents, Oberdeck was certainly not the first to recognize the historical value of never-realized plans for towns and buildings.[18] Moreover, her intriguing stories of Kohler's "paper streets, imaginary blocks, fantasy buildings ... and neighborhoods that remain unrealized" are tightly circumscribed in that they are intended to throw light on the "spatial imaginaries" of architects and residents, and challenge "authorized histories of why materialized spaces came to be." In focusing attention on mining activities, railway building projects, gas-exporting plans, hydroelectric development schemes, and energy transmission lines, Peyton adopts a less stringent definition of the unbuilt. Cassiar, a mine and a town, operated for forty years before it was closed, shuttered, and largely destroyed – or one might say "un-built" – in 1992. The Northwest Transmission Line, treated in Chapter 5, suffered political turbulence and construction delays but was completed eventually, at vastly greater cost than planned. Plans to dam the Iskut and Stikine Rivers to generate electricity (Chapter 3), and schemes for the construction of an LNG plant were unrealized, and both projects were abandoned unbuilt – massive and costly equivalents, perhaps, of Oberdeck's "paper streets." In comparison, the Dease Lake Extension (Chapter 2) might best be characterized as half-built; the projected railroad was never completed, but a railbed was constructed (and served at least in small part the anticipated function of the railway, by improving mobility for humans – and animals).

Not content to simply check off these initiatives as false hopes, shattered dreams, or dismal failures, Peyton aims, admirably, to push the notion of the unbuilt environment beyond Oberdeck's formulation of the idea. Most basically, he uses it to seek "a different angle of vision on economic development and its social and environmental effects" (p. 9). For him, the unbuilt environment idea is an "heuristic device for thinking about what happens in the aftermath of development" initiatives (p. 9). He wants to discover the material traces of these initiatives "on the ground," and he uses the unbuilt notion broadly "to interrogate the altered meanings of nature and place and to understand changes to the relations among nature, local societies, and outside forces," when the public become aware of projected megadevelopment schemes (p. 6). In essence, Peyton would argue that the co-production of resources and the possibilities for their extraction are necessary for resource development to seem possible, logical, and correct. In this respect, the Stikine "has always been a place-in-the-making, constructed by the various discourses and practices brought to it from outside its shifting boundaries" (p. 5).

Important questions spring from this realization: How was the Stikine "made" in relation to other places? How did individual, corporate, bureaucratic, and state actors relate abstract notions of nature and environment to the practical work of mining, hydro-power development, and road-building? What happens when plans go awry? How might development ideas and nascent projects have affected economies and societies, regional ecologies, and the infrastructures connecting northern and southern Canada? Did the skills and expertise brought to and generated in the Stikine by one or another stillborn project persist beyond the duration of that project? How did successive development schemes change "conditions of possibility" in the region?

Peyton explores many of these by attending "equally to the technical and scientific data produced to justify resource development and the altered perceptions of the environment shaped by interactions among newcomers, Stikine residents, and the natural world" (p. 11). The result, he suggests, is a form of layered history, unfolded in five chapters, that examines both the "natural history" (the landscapes and ecologies) of the region and the "geographical imaginations" of those new to it, to demonstrate that "when places are reimagined, material changes follow" (p. 12).

None of the sedimented layers of this story was more marked by material change than the first. Beginning with an evocative recounting of his own visit to the Cassiar townsite in 2010, almost two decades after the mine and mill were shuttered, Peyton carries us past the great, grey-green toad of the tailings pile, through the remnants of a once-thriving town, and into history. There is a haunting quality about this place and this piece of writing. Sixty years before Peyton rode into the valley, it was the site of sporadic amateur prospecting by workers employed on the Alaska Highway to the north; a dozen years after that, it was a major mine, producing over 700,000 tons of asbestos and well over 2 million tons of waste rock each year. Ten years into the new millennium, Peyton discovered that the material remnants of the settlement that grew up around the mine – house lots, waste deposits, abandoned gardens, the rubble where a hockey rink once stood – "exists in outline, but remains tantalizingly ephemeral" (p. 28). Much was sold and removed when the mining company went bankrupt, and much of what remained was flattened. Today these ghostly markers exist amid a scatter of original buildings repurposed as bunkhouses and offices for prospectors, rock-pickers, and a man who makes a living selling old machinery for parts or scrap. "Jade and junk," reflects Peyton, the debris left by forty years of resource extraction, "are Cassiar's commodities now" (p. 32). But memories of Cassiar live on, in a large but poorly organized archive, and in a sprawling, capacious site on the Internet, reminding us of the ways in which places are made and remade, in textual and digital as well as material forms.

Neither Cassiar nor its archival and virtual avatars would have existed without a transportation infrastructure. In the 1950s and 1960s, the Alaska Highway carried Cassiar asbestos to Whitehorse, whence it moved by rail to Skagway for shipment around the Pacific. But the route was long and unsatisfactory. The highway link from Stewart, begun in 1958 and completed in 1972, was built to provide the mine with access to a Canadian seaport. And the Dease Lake Extension, the railroad project that forms the second layer of Peyton's history, was variously seen as a development tool for the region and a way to improve Cassiar's access to markets. Predicated on dreams of resource development, and abandoned in a sea of debt in 1977, seven years after construction began, the Dease Lake Extension was poorly planned and shoddily built. Environmental inventories and impact assessments were hardly thought of. Scarcely monitored and needlessly destructive construction practices exacted a significant toll on local ecosystems; protests from the BC Fish and Wildlife Branch at the contractors' "flagrant disregard" for even the most basic environmental concerns were ignored (p. 75). Still, the abandoned project left 540 kilometres of more-or-less stable, graded roadbed arrowing through areas formerly inaccessible to wheeled vehicles. Hunters and recreationists, as well as mining companies and prospectors, have been quick to take advantage of the opportunities thus presented. As Peyton notes, the effects of the failed Dease Lake Extension project "continue to percolate long after its demise" (p. 87).

When BC Hydro mooted plans to dam the Iskut-Stikine system for electricity generation in 1973, it did so just as the federal government established environmental impact assessment guidelines, and at a time of rising local and provincial interest in the causes of wilderness protection and Indigenous rights. This conjuncture did much to ensure the abandonment of BC Hydro's plans, in 1983, and provides rich material for Peyton's discussion of the ways in which the discourse and debate surrounding this archetypal unbuilt project had "very real material consequences for the landscape and the way people interacted with the environment" (p. 94). First they came to know the Stikine-Iskut environment in much more detail and in ways radically different from before, as BC Hydro produced a sizeable library of scientific and environmental studies about different facets of the area. These framed and codified understanding according to certain principles and, Peyton argues, they worked in some instances at least to marginalize or co-opt various forms of critique. Tahltan people particularly resisted the Western scientific perspectives enshrined in the assessments and the intrusiveness of the surveys on which many of them were based. Disclosing the ways in which people came to know and think differently about the river, Peyton comes to the important and revealing conclusion that it "remained undammed precisely because of the disagreement about what the river means and how it could be best understood" (p. 112).

Although the Western LNG project of the early 1980s, examined in Chapter 4, was also abandoned unbuilt, the present leaches into this discussion more thoroughly than it does into those that precede it, because debate about the prospects for, and wisdom of LNG shipments from, Canada's west coast ports has been front and centre in provincial politics since 2012. So, too, does the present loom large in Peyton's final chapter on the Northwest Transmission Line, proposed as a means to foster development in the northwest in 2004 and revitalized in 2008, then completed in 2014. A major engineering achievement – carrying a 287-kilovolt line across 344 kilometres of rugged, difficult-to-access terrain, by way of almost 1,100 towers – the Northwest Transmission Line has been embroiled in contemporary debate.[19] Just as recent efforts to develop LNG exports have sparked criticism by representing natural gas as a clean energy source, so the transmission line project attracted attention when the federal government provided some $130 million of its $746 million cost from the Green Infrastructure Fund, on the grounds that the new line would allow the Tahltan community of Iskut to abandon diesel generators – although 99 percent of the power brought to the Bob Quinn Lake substation is projected for use by the mining industry. Indeed, one calculation suggests that the number of trees felled and burned in creating the Northwest Transmission Line right-of-way released 120 times more CO_2 into the atmosphere than the annual discharge of all the generators in Iskut.

Each of these sedimented layers provides an intriguing account of an important facet of development in northwestern British Columbia in the last half century or so. *Unbuilt Environments* is impressive for the range of cases it examines, and important for its argument that "the conditions of development – all of the assessments and intrigues, the data and debates, the engineering mistakes and community responses – are critical to understanding how and why environmental change occurs and to detailing the breadth of those changes" (p. 169). We do not have its like in Canada, even though parallel instances of stillborn dreams and arrested designs can be found in full-throated chorus "from Bonavista/ To Vancouver Island/ From the Arctic Circle/ To the Great Lakes waters."[20] Even as it broadens our understanding of the historical, geographical, and economic development of a particular understudied area of northern Canada, this book is much more than a local study. For Peyton, as for the American Edward Hoagland in the 1960s, the Stikine encapsulates both the mythic promise inherent in those alluring, unstable North American destinations, "West" and "North," and the colonial legacies associated with these ever-changing territories.

Unbuilt Environments is also a product of its time, in that it is, most basically, a history of the present. The past of the Stikine is invoked here, not so much for its own intrinsic value as to comment on contemporary circumstances. Peyton does his work cleverly, and he threads an undercurrent of postmodern irony through several of his discussions, but he largely eschews strict narrative chronology for an approach that tacks back and forth through time, to tie past and present together and challenge readers to reflect critically on contemporary circumstances. Hence the strategy, deployed more than occasionally and splendidly displayed in the discussion of Cassiar, of casting the present as a prelude to the past. Hence, too, the interest, evident throughout but again given particularly symptomatic play in the first chapter, in representation and archives and the ways in which environments are catalogued, interpreted, and administered. Ultimately, Peyton's goal is "to interrogate the archives of the Stikine, and to think critically about their many absences and elisions" (p. 166). Following the lead provided by Hugh Raffles's account of Amazonia, Peyton has worked toward an "articulation of natures and histories that works across and against spatial and temporal scale to bring people, places, and the non-human into 'our space' of the present. This is less a history of nature than a way of writing the present as a condensation of multiple natures and their differences."[21]

Concerned with "what-ifs" and "never-weres," and in the way of much recent scholarship, Peyton's *Unbuilt Environments* also shares with Johnston's *Colony of Unrequited Dreams* a certain reluctance "to commit to any particular grand narrative."[22] Yes, each threat of environmental change has spawned "ideas about what northwest British Columbia can look like, what it can produce, and what can be taken from it" (p. 170). And yes, these case studies demonstrate that unbuilt environments have persistent effects. But do these ideas and effects amount to more than thoughts and consequences – to more than the sum of their parts? Peyton only hints at how this might be when he turns, in his final lines, to quote Edward Hoagland: "In the confusion of helicopters and mineral promotions, the question in British Columbia has become the same as everywhere else: How shall we live?"[23]

Historians have often been likened to detectives, forced to identify and correlate fragments of evidence in order to piece together an understanding of what happened in the past. If this characterization has any merit, we might extend it to suggest that Peyton here follows the investigative strategy of Sherlock Holmes, at least as exemplified in Arthur Conan Doyle's "Silver Blaze."[24] This short story, about the baffling theft of a racehorse and the apparent murder of its trainer, is best remembered for the way in which Holmes drew conclusions from a negative fact – finding it curious that no one heard the stable dog barking during the night of the crime. The false starts, delays, and setbacks, the "what-ifs" and "might-have-beens" chronicled by Peyton are the historian's "negative facts." Most scholars, he notes, "focus on the impacts and motivations of projects that have been built; ... we need to look again at those projects that have been put aside, rejected, and cancelled" (p. 7). Considered piecemeal, as separate (un-, partly, or wholly realized) "moments" in the history of the Stikine, individual mines, transportation corridors, power lines, and dams reveal difficult and contested trajectories of development. By bringing these fragments together, and by recognizing the implications of Hoagland's testimony about the helicopters circling over the Stikine in 1968, Peyton turns the Stikine into a synecdoche of the modern world and invites us to think further about the significance of development dreams and their effects on this corner of British Columbia in the last half-century.

As Hoagland framed his question on the banks of the Stikine half a century ago, many British Columbians would have dismissed it as trivial or preposterous. The growth machine had operated at full throttle across the province for some years. Soon after assuming the provincial premiership, in 1952, W.A.C. Bennett famously encountered a curious local on a roadside above the Peace River. The premier asked the trapper what he saw below. "A small, winding, muddy river," came the reply. "Well, my friend," the premier reportedly said, "I see dams. And I see power. And I see development. I see roads, highways, bridges, and growing communities. I see cities – prosperous cities with schools, hospitals and universities. I see beautiful homes with housewives baking bread."[25] Resources paved the road to riches; development in the form of new dams, new mines, and new roads spelled improvement. Some have seen a shift from "a little blacktop government" for the Interior into "a big dam government" for the tycoons in the twenty years of Bennett's Social Credit administration, but there is little question that the province rode a rising tide of economic prosperity through these years.[26] Reflected the premier, "The finest sound in the land is the ringing of cash registers,"[27]

Hoagland stood apart from all of this, of course. He was in northwestern BC on a whimsical journey in search of a vanishing past that he perceived, nostalgically (and inaccurately), as something of a "golden age" of independence and self-reliance. He had come to the Stikine hoping to find the last of Frederick Jackson Turner's stalwart pioneers among the "old men of Telegraph Creek." No wonder he was disturbed by helicopters buzzing overhead. Hoagland's "How shall we live?" question was timely. Others across the continent – and not only those who formed the rising counterculture – were asking much the same thing in the late 1960s – and seeking escape in various places and many ways from what they took to be a too-frenetic, technologically obsessed, and mammon-enslaved society. But by the same token, Hoagland's question was neither new nor ephemeral. It troubled English and American social critics John Ruskin, William Morris, Henry David Thoreau, and others in the nineteenth century, and it was posed repeatedly in the twentieth by thinkers as diverse as E.F. Schumacher and Wendell Berry. It has been asked with increasing urgency in the twenty-first century, by Ramachandra Guha, Clive Hamilton, Bill McKibben, and others.[28]

For two hundred years, an impressive succession of radicals and dissenters uneasy at the course of "progress" have indicted mass consumption, and the rush for resource spoils on which it depends, as the driving forces behind contemporary ills. By and large, critics have seen these developments, running unabated, as robbing life of meaning and desecrating the good earth. In a nutshell, the argument runs, prosperity based on industrial society's exploitation of nature's spoils has produced a spreading epidemic of "affluenza," spawned an irresistible desire for material possessions, and fostered an economic system predicated on an "insatiable" demand for goods. Caught up in this vortex of consumption, people have blithely come to worship false gods. They have forgotten that there is truly "no wealth but life" (John Ruskin), and their foolishness has lent increasing urgency to the question "How much is enough?" (Wendell Berry).[29] As the Northwest Transmission Line promises to unleash a tsunami of development on the Stikine – a wave of change far more powerful and potentially disruptive than any of the economic ripples envisaged or associated with earlier development dreams – such concerns press ever more ominously on the Stikine and its people.

Almost 150 years ago, John Ruskin asked his English readers to imagine private gardens "at the back of your houses … large enough for your children." Then he asked what they might choose to do if they discovered that they could double or quadruple their incomes by "digging a coal shaft in the middle of the lawn, and turning the flower-beds into heaps of coke."[30] Surely, he expected them to conclude, flower beds were better than coke piles. But did Englishmen and women hold this truth only as far as their back fence? Coal shafts and "pit-hills" scarred the once green and pleasant land. Abhor the desecration of their own backyards though they might, Ruskin reminded them, "this is what you are doing with all England. The whole country is but a little garden." In the Stikine, outsiders, rather than local proprietors, threaten(ed) to despoil the metaphorical garden in quest of profits conjured by development dreams. But the larger point of Ruskin's tale remains. The fate of the earth is in our hands and our stewardship of it must extend beyond our individual backyards.

This is the point that Peyton's retrospective works towards, as it comes in its final pages to contemplate what will probably be the first of a substantial succession of mines drawing energy from the Bob Quinn Lake substation, Imperial Metals' Red Chris property on Todagin Mountain. Projected as a massive open pit copper and gold mine processing at least 30,000 tonnes of rock a day for twenty-eight years, this operation might produce over 150 million tonnes of toxic tailings and almost twice as much waste rock. The mountain on which it is located has been described, by Wade Davis, as "a wildlife sanctuary in the sky." To the west is Mount Edziza, sacred mountain of the Tahltan, to the north is the Grand Canyon of the Stikine and to the east are Kablona and Spatsizi. The closest lake, projected to become the site of a tailings deposit, drains into the Iskut River valley – that "garden ... brought alive and given meaning by generation on generation of Indigenous men and women." What good reason is there to defile this space with a huge open-cast mine? Yes, there are rich ore deposits here. But politics and power, influence and avarice, legal systems and property regimes, and particular conceptions of political economy hold more explanatory potential than geology if we would explain why this, rather than any of the other four thousand or so copper properties in the world, should be brought to production – or indeed whether any of them truly need to be developed. "How shall we live?" "How much should a person consume?" Now, as ever, choices reflect values and tell us much about who we are, what we hold dear, and the kind of world that we are making.

Acknowledgments

The publication of an academic book is exciting for many reasons, not least because it provides the opportunity to thank the many people who helped to see it through. This book started at UBC, where I met an extraordinary collection of people in the Department of Geography. Two people in particular deserve special thanks. Matthew Evenden has been an incredible mentor for the better part of ten years. He was the first audience for many of the ideas in these pages and has always offered sharp and incisive comments, pulling back on ideas gone too far and urging more thought on ideas not taken far enough. The lessons I take from my relationship with Matthew – on writing, historical scholarship, professionalism, humility, and work ethic – will guide me for years. I consider myself very fortunate to have benefited from Matthew's intellectual generosity and curiosity. I came to UBC to work with Matthew, but I didn't know that an amazing friendship with Matt Dyce would be part of the bargain as well. We've inadvertently but happily followed each other over the past ten years, all the way to Winnipeg, where we both now teach. Matt's imprints are all over these pages, from the finer points on the idea of unbuilt environments to the scrubbing of the purplest prose. Matt is a great academic partner and even finer friend. Thank you to Matt and Matthew – the book is better for your intellectual engagement, and my life is richer for your friendship.

My time at UBC was shaped by other friendships and intellectual engagements as well. I was very lucky to be there at the same time as John Thistle, Shannon Stunden Bower, Jon Luedee, Max Ritts, Joanna Reid,

Michael Smith, Emily Jane Davis, and Emilie Cameron. Rosemary Collard and Jess Dempsey are true friends and model geographers. Katja Clark, Bryce Firman, Jeremy Redlich, and Gordie Mitchell helped in more ways than they can possibly know. All of these fine folks helped to build the ideas and arguments in the book in one way or another.

The book also owes a huge debt to Graeme Wynn. Graeme's editorial pen is legendary, and this book is greatly improved by the many hours that Graeme spent with it and by the challenges he made to its core ideas. It is a chastening but ultimately rewarding experience, and I thank Graeme for all of the work he put into it. Doug Harris was instrumental in developing the book at an earlier stage, and Liz Harris and Jim Glassman both offered timely critiques at an important stage. Stephen Bocking's insights from afar were instrumental in the shaping of the final product.

At the University of Manitoba, I was very lucky to learn the ropes from Jeff Masuda. Bruce Erickson has taken on Jeff's role as colleague and co-conspirator over the past few years, while Jocelyn Thorpe has also been a source of intellectual support and inspiration since my arrival. I've learned a great deal about professional life from all three. I've also found a wonderful scholarly cohort among the Critical Environments Research Group, whose monthly meetings provide a much-needed intellectual detour from the chaos of term. St. John's College has provided a very comfortable and collegial institutional home. I would also like to express a huge thank you to administrative staff in my departmental home in environment and geography – particularly to Aggie Roberecki and Sam De Pape – for superlative (and very patient) help as I got my feet wet in Winnipeg.

I also received invaluable feedback from numerous colleagues who commented on sections of previous drafts, including Arn Keeling, Norman Halden, Bruce Erickson, John Sinclair, Fabiana Li, Mark Hudson, Cole Harris, and Adele Perry. Tad McIllwraith challenged me with an important question about the contemporary ramifications of the unbuilt environment concept, and I've tried to incorporate his concerns in this book. I've also benefited from the challenges, questions, and insights of conference audiences and fellow panel members at the World Congress of Environmental History 2014, the Canadian Historical Association 2014, the Canadian Association of Geographers 2014, the Association of American Geographers 2011 and 2013, and the American Society of Environmental History 2010.

A host of librarians and archivists have helped over the many years this project was in development. Ramona Rose at the UNBC Archives was immensely helpful in navigating the sprawling Cassiar Asbestos Collection. Jean Ellers Page at the Prince Rupert Archives and the staff at the J.T. Fyles

Natural Resources Library were instrumental in tracking down materials for the Dome Petroleum chapter. I would also like to thank archivists at the Yukon Archives, the BC Archives, and UBC Special Collections. At UBC Press, a very big thanks to Randy Schmidt, who guided the book through the editorial maze. Eric Leinberger made the fantastic maps. I am also happy to acknowledge funding from SSHRC through the Canada Graduate Scholarship Doctoral Program and the Awards to Scholarly Publications Program, and from the University of Manitoba through the University Research Grants Program, the Clayton H. Riddell Faculty of Environment, Earth and Resources Endowment Fund, and the Office of Vice-President (Research and International). A previous version of Chapter 3 appeared in *The Journal of Historical Geography* edited by Felix Driver.

In the Stikine, I would like to thank the many people who shared their time and knowledge with me during several summer visits in Dease Lake, Iskut, and Telegraph Creek. A very special thanks to Jim and Irma Bourquin, whose wit, wisdom, and generosity changed the direction of this work at an important time. Kenny Rabnett amazed me with his commitment to the environment in the Stikine and the Skeena, and with the breadth of his knowledge. In Vancouver, Cynthia Callison and Curtis Rattray have been an invaluable source of information and connection to the legal, social, and political worlds of the Tahltan. Thanks to Sarah de Leeuw for her generosity and hospitality in Prince George. Countless others left their mark on the research. Thank you to all.

Finally, it is a great pleasure to thank my family. My parents – Tony and Mac, Faith and Doug – have been there every step of the way, offering support and encouragement in equal measure. I am grateful to you all. Dad – You've given me an indelible curiosity, a passion for argument, and the conviction to push for what I believe is important and to fight against what I believe is wrong. Mom – You've shown me how to live a life of kindness and generosity and gifted me a love of books that has, in one way or another, manifested in these pages. My brother Toby and my sisters Jenny and Sophie forge their own paths in Calgary and Kelowna. I look up to you all. And to my partner Emma Bonnemaison, for everything, always ... I'm so excited for what comes next.

Unbuilt Environments

INTRODUCTION

The Stikine Watershed and the Unbuilt Environment

The modern history of the Stikine watershed is shaped by a belief in material riches.

Hugh Brody, Stikine: The Great River

THE STIKINE WATERSHED – the traditional territory of the Tahltan First Nation encompassing 52,000 square kilometres in northwestern British Columbia – is under immense and unrelenting development pressure from mineral, metals, and energy companies. In the past decade, the rush to exploit the resources stranded in the bedrock or locked away in tightly compressed shale formations has lured dozens of resource companies and utilities. The area boasts massive mountaintop anthracite coal seams, coalbed methane trapped in the geology beneath ecologically and culturally sensitive landscapes, world-class copper reserves with their own potential tagalong metals economies, water diversions through run-of-river hydro developments, and transportation networks and transmission infrastructure to make it all go. Where these dreams encounter places and peoples, they produce opportunity and hardship in equal measure. The tensions that follow visions of industrial development are nothing new in the Stikine, but they have exacerbated ambivalences in local communities trying to balance the promise of new economic opportunities with the risk of attendant environmental harm and social burdens. Tensions are further amplified because the Stikine is geographically marginal, situated in a remote corner of British Columbia and, for much of the twentieth century, was removed from provincial economic, political, and cultural currents due to confounding issues of access, infrastructure, and investment. Generally, opportunities within industrial capitalist economies were circumscribed, and lifeways were partially defined through a set of predetermined environmental possibilities. This marginality is a marker of distinction for many; it draws entrepreneurs with visions of opportunity and allows residents to maintain a place apart.

Dreams of development in places like the Stikine reveal an ambivalent relationship with nature. Consider the conflicted position of conservationist Tommy Walker, who described the Stikine as a place of pristine wilderness imperiled by the onslaught of "the industrial octopus."[1] He arrived in the area in the late 1940s to start an outfitting business and fulfill his dreams of economic independence. By 1960 he was writing to a colleague,

> I am afraid we have seen the best of the north country. From now on changes in methods of travel and communication will make access so much easier and there will not be any real wilderness left. It is just one of those things that have to happen. I just love a wild uninhabited country and I am so thankful that I have known some of our wilder areas before they become ruined by the encroachment of civilization.[2]

The tension between dreams of wealth and nostalgia for a world he had helped to destroy is threaded through Walker's archive. Although he dedicated much of his life to preserving the Spatsizi Plateau as a park, he never questioned the dispossession of those who had previously used the land on which his outfitting business depended.[3] Similarly, aspiring novelist Edward Hoagland, seeking an antidote to the stultifying effects of his New York City life, came to northwest British Columbia in 1966 on a naturalist's whim, "because people will want to know what these wild places were like when there are no more wild places."[4] Fifty years later, leading public advocates are saying essentially the same thing about the northwest. As David Suzuki has lamented, "This beautiful and pristine wilderness area is under industrial threat … that is occurring in a vacuum of silence."[5]

The Stikine, like many places at the margins, has often been the focus of the romance and nostalgia associated with "places left behind." The allure of an authentic "wilderness," the pristine beauty of a rugged, unforgiving landscape, the possibilities of untapped resources and the curious habits of its residents have drawn adventurers, prospectors, ethnographers, and settlers to the region. From John Muir scrabbling over its glaciers in the 1870s to Hoagland's journey upriver to meet the "Old Men of Telegraph Creek" in the late 1960s, from Warburton Pike's ambitious rail and mining schemes at the turn of the century to Royal Dutch Shell's recent attempts to harness coalbed methane concealed beneath the "Sacred Headwaters," the Stikine has seen numerous interlopers, all searching for a particular experience or enterprise embodied in the river and its surrounding plateau.[6] It has always been a place-in-the-making, constructed by the various discourses and practices brought to it from outside its shifting boundaries. But how, exactly, was the Stikine "made" in relation to other places? How did the harnessing and exploitation of resources shape both the development of the Stikine and its connections to other places? How did individual, corporate, bureaucratic, and state actors relate abstract notions of nature and environment to the practical work of mining, hydro-power development, and road-building? How did these things affect the cultural understandings of the area's original inhabitants?

Another more populist rendering of the Stikine frames the region as a geographical legacy of grand dreams and even grander failures. Many accounts of northwest British Columbia make much of the rhetorical use of "failure" and "marginality" in discussing problems of and prospects for development in the region. Much like American journalist Joel Connelly, visitors have tended to see the Stikine as "a land of big dreams, and at times, big failure."[7] John Faustmann, with a wink to the region's gold-mining history, interprets northwest British Columbia as a place that has seen "a string of big plans that never panned out."[8] Ethnobotanist and National Geographic explorer Wade Davis, who perhaps more than anyone has sounded the alarms about "the tsunami" of industrial development awaiting the Stikine, claims to be almost "numb to the endless series of grandiose, if ill-conceived megadevelopment plans" that have been brought to the Stikine from outside.[9] Many locals employ the same trope to explain the Stikine as a place apart. In a much-cited popular periodical from the late 1970s, we learn that "the story of the Stikine has essentially been one of false hopes and would-be developments ... of railway lines not quite constructed; grand telegraph proposals that failed (and one that succeeded if only for a few decades); gold mines that petered out; copper and coal deposits that defied development ... and on and on."[10]

A sense of frustrated progress resounds through accounts of the region and its distinctiveness. Explorer and entrepreneur Warburton Pike offered a particularly egregious example early in the twentieth century: "apart from minerals this country is absolutely worthless."[11] Repeated over the years, such opinions consolidated "the story of the Stikine" as a self-consciously failed landscape removed from modern standards of industrial progress. Failure in this instance is contingent on an understanding of marginality created by acclamation rather than being based in any biophysical reality. As Bonnie Demerjian has it in her popular account of the river and its environs, "The inaccessibility has always challenged those who wished to exploit its riches."[12] I take these claims of failure seriously to see what emerges from unsuccessful development projects and unrealized infrastructure programs.

To understand why the Stikine is so often regarded as a graveyard of development dreams, this book examines evolving patterns in interaction between the physical place and human societies anxious to represent, understand, and use an out-of-the-way area increasingly open to outside influences. Plans for development begin in people's imaginations. Indeed the area was mostly unfamiliar to the state – and known (other than to its Indigenous inhabitants) to only a few hundred prospectors, traders, and missionaries until the discovery of gold in the Yukon in 1896. By popular reckoning, the area has remained in the shadows. To borrow a phrase from Julie Cruikshank, it is a place of "imaginative possibilities."[13] Yet, like any other place, the Stikine is complicated by the memories and the meanings and representation that people ascribe to it. Place-making here was a crucial site of political contest and of the production of cultural meaning. The interconnections between the complex concerns of colonialism, ecology, and the movements of capital make the Stikine an important site of historical and environmental inquiry. This book is a historical geography of the failure of infrastructure development and extractive potential in northwest British Columbia. I look at railway projects, gas export, hydroelectric development, mining, and energy transmission to interrogate the altered meanings of nature and place and to understand changes to the relations among nature, local societies, and outside forces.

Visions of a northern landscape dotted with dams, railroads, transmission lines, and mineral projects have inspired industrial entrepreneurs to conceive countless schemes for the extraction, harnessing, and transport of goods and energy from the north. These schemes have transformed the environments, economies, and social lives of northerners. But for every successful megaproject, there are dozens that survive only on paper, abandoned before completion, or that operated only for a short time. These visionary projects can be conceptualized as unbuilt environments, a term that signals the environmental and social side effects of planned but unrealized megaprojects that were conceived as development schemes, lucrative extractive economies, or smaller-scale sustainable resource economies. I develop the unbuilt environment concept to question the social meaning and environmental effects of resource development failure in the Stikine watershed.[14]

Unbuilt environments can be recovered through archival evidence, material remains, and a careful appraisal of both altered perspectives of

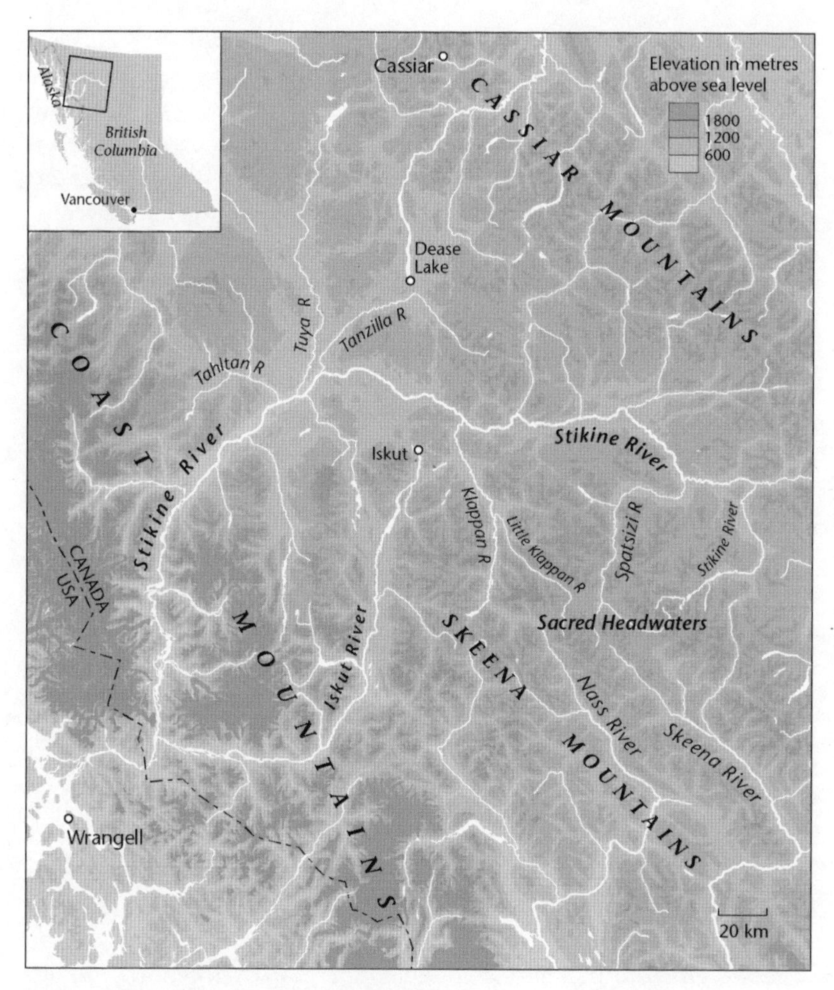

Figure I The Stikine. Map by Eric Leinberger.

nature and the shifting boundaries of human–nature relationships. Most scholars justly focus on the impacts and motivations of projects that have been built; however, we need to look again at those projects that have been put aside, rejected, or cancelled. By doing so, we gain a deeper understanding of the relationship between development ideas, nature, and the north in Canada. Such projects have lasting effects on the society and economy of the north, on regional ecologies, and on the infrastructures connecting north and south. Through an analysis of extractive economies and megaprojects that did not materialize or were only partially completed, this book furthers our understanding of the historical, geographical, and economic development of an understudied area of northern Canada. I contribute to a re-evaluation of nature and the built environment by examining the planning of, debate about, and scientific engagement with northern geographies to understand better how newcomers and residents knew nature in northwest British Columbia.[15] In many ways, the region's present economies, environments, and social relations are influenced by the legacy of unbuilt or incomplete infrastructure megaproject dreams brought to the area in the name of progress, improvement, and industrial profit. I examine various phases of development over the postwar period and ask what happens when plans go awry. What are the unintended outcomes? How do the remains of one development process or project influence later schemes? Answers to these questions highlight the conflicts, tensions, and contestations that follow from the ambitions, calculations, assessments, and failures of developers pushing at the boundary of the industrial frontier.

A large literature on the state and northern development in Canada emphasizes the historical particularities of individual development episodes and the uneven geographical development that follows the resource cycle of boom and bust.[16] In general, this literature highlights the economic exposure of northern resource peripheries to the changing patterns and processes of supply and demand in southern metropolitan economies and the role of the state and corporations in framing development decisions. While the idea of the unbuilt environment broadly parallels these concerns, it focuses less on regional economic development than on phases of transportation infrastructure development, energy, and mining – cases which, in the Stikine, do not represent uneven development so much as unrealized corporate and state projects.

Economic historians have used the notion of path dependence in discussions of development at the margins to suggest that development is circumscribed by previous decisions that structure and limit possibilities. The concept is useful for thinking about how the artefacts and material traces of previous development regimes influence the next iteration of development.[17] My approach borrows from the iterative notion at the heart of this historical analysis but, rather than focusing on path dependencies, it examines the possibilities that follow from consecutive efforts to develop the Stikine. By concentrating on the expertise, knowledge, and management brought to and generated in the Stikine, this study broadens the claims of the path dependence framework to suggest the need to focus on the capacities that are created by failed schemes. It also extends the idea of "material traces," employed by William Turkel to show how people read the past in historical landscapes through artefacts, providing a way to read the environment as a contested archive of historical change.[18] This book widens this analytical framework to include a study of the discourses that can be tracked from the things left behind by unsuccessful development efforts.

I examine a series of encounters in time and space in case studies extending from mining projects and railroad construction to hydroelectric development and hydrocarbon export. The cases reveal a range of outcomes from construction and abandonment, through partial completion and incompletion, to dreams unrealized. The unbuilt environment concept thus provides a way to think about how the processes of unfulfilled development schemes bear effects. In this region, development projects have been often unrealized, partially built, or operated within relatively short time horizons. Some have operated successfully for a period of time, only to be shut down by external economic factors. Because the historical geography in the Stikine has exemplified constrained ambitions and frustrated developments, side effects and outcomes take on additional importance in a study of social relations and the environment. Examining the Stikine with an eye to outcomes and side effects raises questions about the particular legacies of development in a peripheral environment where extractive economies have been enormously important and where sustaining them has been difficult. The idea of the unbuilt environment is thus not a predictive theory or a literal description but, rather, one that offers a different angle of vision on economic development and its social and environmental effects – a heuristic device for thinking about what happens in the aftermath of development. The concept outlines the co-production of resources and extractive space within which resource development has been made to seem possible, logical, and correct. Through an analysis of megaprojects and development schemes that fell short of the aspirations embedded in their descriptions, I frame the Stikine's historical production as a site of "development."

A broader northern development context in the postwar years positioned megaprojects as the primary method for opening up the vast northern realm of the continent. There are many examples of ambitious development schemes across subarctic and Arctic North America; many of these have faltered or failed. In this sense, the Stikine is not unique but, rather, a location characteristic of prevailing development wisdom in northern Canada. A paradigmatic example is the Mackenzie Valley Pipeline proposed in the 1970s to deliver Arctic gas to northern Alberta.[19] The pipeline was never built, but it galvanized local opposition on social, environmental, and economic questions and propelled northern resource development onto the national stage as never before. Other examples range from the Alcan or Alaska Highway to the Northwest Staging Route, from the construction of Inuvik to the myriad mines that dot the subarctic landscape, from the Canol pipeline from Norman Wells to Whitehorse and beyond, to the many transportation routes constructed through John Diefenbaker's "Roads to Resources" program. Lacking infrastructure and economies of scale, northern development has always been subject to the big dreams of entrepreneurs, speculators, and governments eager to tap new resource veins so that they might prosper in the global economy.[20]

This developmental agenda spread even farther afield. In Alaska, the Rampart Dam project, originally conceived as a 5,000-megawatt-generating dam on the Yukon River in the early 1950s, was abandoned in 1970 amid acrimonious technical and bureaucratic in-fighting and environmental protest. Rampart would have created the world's largest human-made reservoir at 28,100 square kilometres, displaced 1,500 people, blocked salmon passage, and flooded important wetlands.[21] At much the same time, the Yukon government sought to direct the Yukon River into the Taku River, and thus the Pacific Ocean, while Alaska proposed a competing strategy to divert the same river into the Pacific by channeling it into the Taiya River.[22] All three projects were part of larger plans to attract aluminum smelters to the north. In some cases, like these competing river diversion schemes, effects were geopolitical as much as social and environmental.

Similar dynamics played out in the "provincial norths," an interpretive model of development that formalized the analysis of the internal colonial relations between south and north as they played out in a provincial context.[23] Development megaprojects were also preeminent within this paradigm, with mining and hydroelectricity front and centre.[24] Once again, examples abound. Isolated mining economies popped up in places such as Uranium City, SK; Elliot Lake, ON; Flin Flon, MB; and Tumbler Ridge, BC. Northern provincial landscapes, habitats, and homelands were submerged by mega-scale hydro projects such as the multi-stage James Bay Project in northern Quebec, the Churchill River Diversion in central Manitoba, and the W.A.C. Bennett Dam on the Peace River in northeast British Columbia. This kind of breathless industrial development yielded impressive results, but it also required that innumerable schemes be abandoned before launch.

Such is the broader context for an examination of unbuilt environments in the Stikine, which contains two equally important objects of analysis.

The discursive effects of megaproject and infrastructure development in the Stikine show how interested parties engaged with and invoked the river and the natural environment as they debated the terms of progress and industrial improvement. Equally, the material traces left by the work on the ground – such as road construction and transmission lines – show the effects of rapidly accelerating scientific engagement. Discourse relating to development not only changed environmental perception in the Stikine but also had an impact on the physical environment by increasing human, scientific, and technological engagement in the region. The goal is to build an enlarged sense of historical possibilities that can speak to the actual physical interventions into the landscapes and attend to the ideas and potential displacements that emerge out of them.

Failure and Effects

Development schemes in the Stikine emerge in a context that lays bare the hubris and myopia that often characterize resource mobilization. By maintaining an analytical focus on the conditions of possibility that emerge from each development scheme, I attend equally to the technical and scientific data produced to justify resource development and the altered perceptions of the environment shaped by interactions among newcomers, Stikine residents, and the natural world. This new analytical framework relies upon and connects earlier work with a normative and analytical concern for the unequal effects of human–environment relations, drawn from political ecology (or critical resource geography), regional historical geography, and environmental history in Canada.

I follow a long tradition in regional historical geography scholarship that seeks to represent a place through fine-grained analysis of its salient features and borrow analytical techniques from environmental history to complicate the natural histories of particular regions. Material history also looms large in this work.[25] Recognizing that study and area boundaries are developed in different ways and that place is continually remade, contested, and negotiated, my approach renders the traditional metrics of geographical inquiry – scale, nature, landscape, environment, place – into malleable concepts acquiescent to history, politics, and environmental dislocation. Places such as the Stikine – much like Bill Turkel's Chilcotin, Claire Campbell's Georgian Bay, or Jocelyn Thorpe's Temagami – are perpetually reimagined as historically contingent but ultimately new places, formed and reformed.

When places are reimagined, material changes follow. A broad cross-disciplinary literature within the field of political ecology analyzes "local conflicts over access to resources that originate with colonialism and the expansion of global capital."[26] The major themes of political ecology – resource access, environmental governance, the application of scientific and technological management, and environmental justice – are useful for dealing with questions of resource politics, especially when combined with the rigorous empirical and archival methods of environmental historians. Given their analytical and normative concern with the legacies of colonialism, political ecologists originally focused on social and environmental inequities in the "global south" under conditions of unequal economic and social relations.[27] The discourse of improvement and the push for progress have been at the centre of this critique of embedded inequality.[28] Much of this work treats the "friction" of post-colonial interaction. As conceived by Anna Tsing, friction builds out from the messy, unrelenting dynamism of power, knowledge, and hegemonic arrangements and exhibits them in the distance between the state, its subjects, and a historicized nature.[29] Within this frame, analyses of development in action show the compromises and contestations that often undermine the benevolent intentions of those who would improve the lives and lifeways of those in the global south. Development is not taken as a pre-determined analytical construct but, rather, as an object of inquiry. Places are unbounded, fragmentary, and fluid. The multiple meanings ascribed to nature within this loose structure are further complicated as they are analyzed as objects of new, emerging power relations. The most compelling work, like that of Tania Murray Li, shows that the "will to improve" has a long history and often functions as a project of rule, while simultaneously working to build infrastructure, human capital, and economic capacities.[30] In this book, "the state" refers to a complex of agencies, institutions, government ministries, crown corporations, and the people who worked for them, that sought to develop and mobilize the resources of northwest British Columbia and, as a result, governed possibilities for the improvement of the area.

By conceiving of history as "sedimented" a new generation of scholars has attempted to hold in hand both the "geographical imaginations" of those new to a region and a "natural history" that exists in landscape formations and the changing ecologies of rivers, people, and animals. Hugh Raffles writes this layered "natural history," which he sees as an "articulation of natures and histories that works across and against spatial and temporal scale to bring people, places, and the non-human into 'our space' of the present. This is less a history of nature than a way of writing the present as a condensation of multiple natures and their differences."[31] In the postwar period, the idea and practice of development has worked as an important concept and reality in the Stikine. Results have been decidedly mixed, both for inhabitants and the development dreams. Through five case studies, I invoke the multiple natures evident in the Stikine during the better part of sixty years of contest and conflict over resources.

Political ecology destabilizes bounded notions of place but it is always indebted to specific geographies of difference. Over the past decade, there has been a concerted effort to bring the analytical mandate of political ecology to the "global north" in particular, to focus on the environmental dislocations experienced by marginalized peoples and spaces in more industrialized societies.[32] Sometimes identified as critical resource geography, this broad framework is attuned, in the words of Gavin Bridge, to "the processes by which particular parts of the environment become produced as resources. Such an approach argues that producing resources for capitalist exchange ... requires the regulatory practices of science, technology, capital and law in order to rework existing natures and space."[33]

Resources, then, are historical and social and as much as biophysical fact.[34] Canadian environmental historical scholars have been less explicit with these connections, although the appraisal of resources as a set of social relations has become a foundational interpretive framework for many of these same scholars.[35] This has been particularly true for historians and geographers working on northern landscapes. Some lean towards material outcomes in their analyses. For Arn Keeling and John Sandlos, for instance, "the landscapes produced by industrial development in the North ... became material expressions of the power and priorities of southern Canadian institutions, technologies, and ideologies,"[36] while for Liza Piper an emphasis on history reifies the "cumulative impacts of successive exploitation" of northern environments.[37] Others focus on the conditional effects of culture and discourse. For Caroline Desbiens, the goal is "to analyze the work of culture in laying out paths of economic development" while maintaining, as I do, that "discursive formations have material impacts" that can be traced in the landscape of northern Quebec.[38] Similarly, Emilie Cameron assesses the stories told about and through the history of Kugluk "to both *place* and *displace* [outsider] relations with the contemporary North."[39] This can all be seen alongside Nancy Lee Peluso's call for a critical socio-natural history that positions the commodification of nature as both a historical and political process, contingent in large part on the paradigmatic institutions of the modern resource state and of the colonial context from which it springs.[40] I borrow from this historicized sensibility in my analysis of environment and resource conflict in northwest British Columbia while being mindful of the multiplicity of natures, porous boundaries, and human histories of use and industrial enterprise.[41]

Ambition underpins development dreams. In the Stikine, the notion that ambitions have been mostly frustrated has been reinforced by popular commentators looking for a narrative hook and embraced by developers eager to imply that their project would be the one to turn the tide of progress. Yet these dreams remained unfulfilled. The cases under consideration are never about a failure of imagination but, rather, about breakdowns and disjunctures in execution. Following scholars at the leading edge of this research at the intersection of innovation, enterprise, and failure, I ask questions about how failures influence the growth of geographical knowledge.[42] This is crucial for the Stikine, where industrial, scientific, and state-sponsored professionals, eager to accumulate and disseminate knowledge, entered the region alongside and as part of the failed development schemes that form the case studies of this book.

Development failures produce their own conditions of possibility by creating a perception of the Stikine as a part of the province where resources are located and can be mobilized, with careful planning and initiative. In this sense, failure is future-oriented and productive of both the prospect of industrial extractive economies and the resource conflicts and contestations that accompany mines, dams, transmission lines, and energy projects. Failure, conceptually and analytically, looms large in this book. We might call this a political ecology of failure in the sense that failure is both an outcome and product of resource conflict, a question of access and accountability, and the complex politics embodied within the project of environmental governance.

Two books yield critical insights into the nature of failure and its aftermaths. In *Seeing Like a State,* James Scott is concerned with developing a diagnosis of failure in order to question why it happens and to identify the disorders of governance that make failure possible, if not inevitable. In his "account of the logic" behind development failures, large-scale social and environmental engineering projects are unsuccessful because they fail to recognize local knowledge.[43] *Unbuilt Environments* builds on Scott's fundamental questions, but it is less concerned with the reasons for failure than with the effects of failure. The question is not why did something fail but, rather, what did failure produce, what kinds of effects did it have?

In pursuing this path, I draw from a work that preceded Scott's by nearly a decade. In *The Anti-Politics Machine,* James Ferguson describes the failure of an agriculture improvement scheme in Lesotho but maintains that, precisely because of its failure, it "had a powerful and far-reaching impact" on the region.[44] The construction of roads, the availability of government services, and the creation of a local administrative centre among other speculations suggest that "what is most important about a 'development' project is not so much what it fails to do but what it does do: it may be that the real importance in the end lies in its 'side effects.'"[45] Scott's focus on failure and Ferguson's focus on the effects of failure have been sharpened even further by Tania Murray Li, who asks, "What do these schemes do? What are their messy, contradictory, multilayered, conjunctural effects?"[46] The task ahead, then, is to ask what was left over in the wake of the uncoordinated and often incoherent development projects brought to the Stikine and to examine the durability and historical contingency of those effects.[47]

The stories I tell about the use of nature in the Stikine are multifaceted and incomplete. Indeed, part of the point is that they bleed into each other and leave traces in the historical imagination of the region, which can be read in the projects that follow. These are the material and discursive "unintended consequences," an analytical mode that has found purchase in both political ecology and environmental history.[48] Failure and success both entangle similar processes – whatever the outcome, scientists and surveyors come north to enumerate; bureaucrats project ideal scenarios; locals, advocates, and opponents mobilize debate and opinion on matters of development. In all cases, the environment is the object of transformation. As Richard White argues in his monumental history of American railroads, industrial dreamers "created modernity as much by their failure as by their success."[49]

Stikine Outlines

Though it lies just beyond the boundaries of the Stikine watershed, the Cassiar – the subject of my first chapter – was long the economic and service centre of the Stikine region. Opened in the early 1950s, Cassiar was active for forty years until shaky financial management, fluctuating asbestos markets, and engineering difficulties forced closure of the mine, and the adjacent company town, in 1992. This chapter visits three sites – the abandoned townsite, the tailings pile, and the pit and mill where the outlines of community and industry can still be perceived; the archives at the University of Northern British Columbia, where the town's remaining material history resides in several thousand boxes; and the virtual town, recreated online by former residents whose connection to the place persists in spite of the town's erasure – to provide details of the mine, town, and company and to outline the eventual failure of the company. Cassiar and the way of life it sustained were dismantled as a result of corporate decisions, consigned to a footnote in British Columbia's capricious mining history. Yet within the stories of failure and undoing, there are smaller examples of planning, management, calculation, and corporate social responsibility that touch directly on the side effects of unrealized development dreams. The ongoing environmental ramifications of closure and abandonment in Cassiar raise questions about the connections between nature in a marginal place and the far-reaching impacts of global cultural and economic forces.

The development of transportation infrastructure was a perennial problem for the citizens and corporate leaders of Cassiar. Throughout the twentieth century, the movement of goods, commodities, and people was a vexing question in northwest British Columbia, given the sparse and poorly maintained road and rail infrastructure. Successive BC governments built the Dease Lake Extension to "open up" the region to mineral and metals development by providing a link to the BC Rail network and to an open-water seaport at Stewart. During the 1970s, BC Rail contractors laid a gravel rail bed from Fort St. James to Dease Lake, a distance of over five hundred kilometres. Spiraling cost over-runs, poor environmental stewardship, and lax engineering and assessment standards produced a maelstrom of criticism. By 1977, the Dease Lake Extension was in jeopardy, and, after recommendations by the Royal Commission on the British Columbia Railway, the project was abandoned by BC Rail and the provincial government. The rail bed remains, a legacy of the failure. Chapter 2 interrogates the environmental assessment initiatives, engineering, and planning mechanisms of the Dease Lake Extension project, while paying particular attention to the environmental legacies of failure.

As the Dease Lake Extension's drive north was drawing to a close, BC Hydro initiated an ambitious hydroelectric generation program. Chapter 3 highlights the tensions surrounding the company's proposal to build five dams on the Stikine and Iskut Rivers. These would have markedly improved the grid capacity of BC Hydro while creating large impoundments, disrupting riparian ecosystems, and interrupting fisheries practices. To demonstrate the environmental, economic, and social feasibility of the damming project, BC Hydro embarked on an ambitious assessment program designed to catalogue, enumerate, and evaluate natural and human resources in the watershed. I analyze the assessment apparatus to look at how the collection of new scientific, technological, and ecological data affected perceptions of nature in the Stikine. In addition, I consider the effects on peoples' relationships with nature by analyzing responses to the new categorization of the environment, the assessment of attributes of nature and how nature might be valued by communities, and the implementation of research strategies around the dams. The episode marked the first real emergence of a "corporate ecology" in the Stikine, where corporate interests mediated the tensions around resource use and local livelihoods were subject to technocratic decision-making processes at BC Hydro.

Concurrently, another aspirational energy economy was being proposed farther south. In the early 1980s, Dome Petroleum of Calgary developed a plan to build a liquefied natural gas (LNG) processing facility at Grassy Point on the central coast of British Columbia to export gas under contract to Japan. A network of pipelines would connect Grassy Point, adjacent to the First Nations community of Lax Kw'alaams, to Dome's pioneering gas-extraction scheme in the Beaufort Sea. Under Dome's Western LNG Project, as it came to be known, Arctic gas would be piped to the Pacific Coast, frozen, liquefied, drastically reduced in volume, and then shipped to Japanese cities. Although the project was never completed, the particular conditions of possibility that it generated highlight the persistence of energy economies, a point that has been brought into sharp focus by the contemporary scramble to export gas from Grassy Point, Kitimat, and neighbouring locations along the coast. In Chapter 4, I argue that Dome's Western LNG Project required manipulations of both the materiality and the meaning of gas in the effort to move the product across the Pacific Ocean. In a sense, this chapter is an outlier, at least geographically, in that it pushes beyond the boundaries of the Stikine watershed, but also because of the contemporary resonance of LNG dreams. In addition, Dome's sightlines were angled towards Japan, rather than inwardly or to domestic markets. Yet the analytical concerns are consistent with other cases, and it provides a compelling test of the methods of this approach in a neighbouring sub-region of northwest British Columbia.

The final chapter analyzes a contemporary resource and infrastructure initiative, the mundane nature of which belies its intended investment-galvanizing qualities. The Northwest Transmission Line (NTL) brings power from the provincial grid to the northwest and will eventually extend all the way to Dease Lake. Proposed under the "green energy" banner, the NTL is meant to serve an eager mining sector that needs access to new, cheap power for its mines, including several very large copper and coal properties in various advanced stages of the environmental impact assessment process in northwest British Columbia. All of these properties require assurance that necessary infrastructure will be in place if they are to proceed with development. I consider the provincial government's rationale for the NTL, as well as the BC Environmental Assessment Office evaluation of the NTL, drawing connections between the social and environmental lives of northwest BC residents and the mining companies that seek to operate in their midst. Sources for this discussion are very recent, culled mainly from the popular press and from malleable and constantly updated government and industry documents. This chapter develops, along with much of the book, an argument that the environmental impact assessment process prioritizes description and approval over protection and oversight.

The development motivations behind the projects discussed in each chapter have contemporary iterations that promise new extractive frontiers. This is particularly the case in the final two chapters, which speak directly to development narratives that are still unfolding. Copper and coal mines, energy generation and transmission schemes, and infrastructure programs are on the verge of realization. I argue, however, that the current resource mobilization is predicated in large part on the conditions of possibility established by previous attempts to extract, harness, and transport the "stranded resources" that have, up to this point, eluded development dreamers and industrial pioneers. This book shows how five resource development encounters, none of which lived up to developers' expectations, embedded the perception that northwest British Columbia is a promising resource space. The contemporary stories of LNG and energy transmission are, in many respects, contingent on the resource encounters that came before. Here we confront the layered effects of successive development projects to show how the reimagination of past resource landscapes shapes contemporary debate around the politics and possibilities of development.

The Stikine, sometimes referred to as Cassiar, is Tahltan Country. Although the focus of this study is on more or less incomplete projects from the late twentieth century, this does not imply an absence of Tahltan social, cultural, and economic presence throughout the Stikine watershed, or anywhere they claim as their ancestral homeland.[50] The Tahltan, and the Lax Kw'alaams farther south, have occupied this land for thousands of years. This book chronicles the history of the region in the years after the Second World War but, of course, the preceding centuries assert other preconditions of possibility on the environment.

Perhaps the clearest way to illustrate the long history of Indigenous settlement and resources use in these areas is to reference an extraordinary document issued by the Tahltan in the white heat of an ambitious Indigenous rights campaign at the beginning of the twentieth century.[51] The 1910 Tahltan Declaration asserted the "sovereign right to all the country of our tribe," which, like most of British Columbia, had (and has) never been ceded to the Crown, through treaty or otherwise. The declaration, signed by eighty-three Tahltan individuals and brought south by ethnographer and big-game hunting guide James Teit, affirmed that the Tahltan held their territory intact "because our lives depended on our country."[52] The rights, values, and lifeways asserted in the text of this document forms the legal imperative and normative basis for all Tahltan interactions with (and their claims to territory and the resources of) the Stikine. The 1910 Tahltan Declaration reinforces long-standing Indigenous rights, but it can be projected forward in the face of the many development dreams brought to the Stikine over the next century.

The Stikine watershed is the ancestral homeland of the Tahltan, who have lived in and used the watershed "since time immemorial," a phrase adopted by many First Nations to invoke the indefinite depth of pre-contact history in British Columbia, beyond the reach of conventional memory of the kind held in state archives. An Athapaskan-speaking, semi-nomadic hunting and fishing people, the Tahltan travelled between temporary settlements as the pursuit of resources and subsistence dictated. Tahltan families would gather in the summer near the Stikine River to fish, feast, and trade with neighbouring groups: the Tlingit to the west and the Kaska and Carrier-Sekani peoples to the east and south. Animosity, periodically resulting in conflict, characterized relations with the Taku Tlingit to the north and the Nisga'a to the south. Relationships were fluid, and borders overlapped. Summer abundance gave way to fall and winter privations when small family units would disperse to hunt on the plateaus and grasslands above the river. Allegiance to these family groups, divided broadly into Wolf and Raven/Crow phratries, determined who had the right to hunt, fish, and live in particular parts of the watershed.[53]

Tahltan ethnographers suggest that the group's first contact with Europeans was with Russian traders in 1799, although British explorers James Cook and George Vancouver and French explorer Jean-Francois de La

Pérouse all sailed near the mouth of the Stikine River before the close of the eighteenth century.[54] The Russians founded a trading post on the site of present-day Wrangell in the Stikine delta in the 1830s. The outpost was beset by a series of closures, hardships, and territorial disputes in the decades that followed. Uncertainty about property and ownership continued until the Alaska–Canada border was firmly settled in 1903. But it was not only newcomers who quarreled over ownership. The Stikine Tlingit were important trading partners of the Tahltan, but periodic wars over territory and river-related resources also characterized their inter-tribal histories. Tahltan scholar and professor of journalism Candis Callison notes that today's Canada–US border closely approximates the non-state border established between the Tahltan and the Tlingit of southeast Alaska.[55]

Further inland, contact was less frequent though it was also facilitated by trade. The Tahltan often acted as intermediaries between European traders and their neighbours. In 1838, Robert Campbell, an agent of the Hudson's Bay Company (HBC), met a group of Tahltan at present-day Telegraph Creek. The HBC wanted to set up a post on the Stikine, and the Tahltan brokered a deal between Campbell and the Tlingit, who were also present at the meeting.[56] Fearing for his safety, Campbell soon left Tahltan territory. As a result, a full-time HBC post was not established until gold was discovered on the banks of the lower Stikine in 1861. Another small rush ensued in 1874 when gold was discovered near Dease Lake. Drawn by potential riches, for the first time, a small population of non-Tahltan newcomers lived year-round in the watershed.[57] Small, localized placer-mining activities have persisted ever since. Representatives of state and church institutions followed the miners, and merchants arrived to supply the material needs of itinerant miners, overburdened bureaucrats, missionaries, and, increasingly, the Tahltan themselves.

Traditional trading networks were frayed by the influx of goods and new markets around manufactured goods. Semi-permanent villages were established around trading posts. Yet travel between these villages, as well as contact with newcomers, Wade Davis contends, facilitated the transmission of new pathogens that exerted their own terrible power on Tahltan bodies and families.[58] The introduction of new diseases caused great hardship and deepened the cultural dislocation consequent upon new trade relationships. Small pox outbreaks in 1832, 1847–49, and in the 1860s, typhus in 1918–19, measles in 1920 and again in the 1940s reduced populations and harmed Tahltan economic, cultural, and social stability.[59] Population losses may have reached as high as 90 percent.[60] Miners brought goods alongside new techniques for opening up the watershed. Steamboat service began in the 1860s and lasted, in one form or another, until 1969, when floatplanes and helicopters finally displaced river traffic as the primary mode of transport.[61] Modern communication technology also left its mark. The Western Union Telegraph Company failed in its attempt to lay a cable from San Francisco to Moscow, overland across western North America, across the Bering Strait, and through Siberia. Engineers working on the Collins Overland Telegraph, as the project was known, raced to be the first to lay cable across the Pacific Ocean. The bold project was abandoned after a competing company outdid them by laying an underwater cable between Newfoundland and Ireland, but the episode gave the only settlement on the river its name, when several hundred tonnes of telegraph cable were abandoned on the banks of the river at what is now Telegraph Creek.[62]

The Klondike gold rush, far beyond the borders of the Stikine in Yukon Territory, inspired another ambitious telegraph scheme. The Yukon Telegraph, completed in the early years of the twentieth century, ran through the Stikine until it was closed in the mid-1930s.[63] As many as five thousand prospective miners came through northwestern British Columbia during the rush north to Dawson City between 1896 and 1899, but only a few hundred made it through to the gold fields over the "All-Canadian Route" through the Stikine (not to be confused with the route out of Edmonton bearing the same name).[64] A rail connection from the Stikine north to Teslin Lake, promised by the Canadian federal government, was never built. Instead, a derisory trail was hastily broken. Telegraph Creek and Glenora, twenty kilometres downriver, flourished briefly, only to sink again into torpor.[65] However, a substantial hunting and guiding economy, providing access to the area's abundant wildlife, thrived during the first decades of the century and remains a backbone of the northwestern economy to this day, providing a substantial income to Tahltan and settlers alike.[66] All of these events reconfigured Tahltan social relationships, ushered in new economic opportunities, and inaugurated competing notions of land use and resources, as did each of the developments discussed in the following chapters.

In an essay about a mostly forgotten mining settlement in central Alaska, William Cronon unearths "the paths in and out of town" to highlight how seemingly marginal places are intimately connected to the outside world through industrial networks, consumption practices, and changing relationships with nature.[67] Cronon's discussion of Kennecott, copper mine and boomtown, rationalizes the movements of goods, capital, and energy through developing corporate agendas and human–nature interactions.

In an oblique way, Cronon deals with the unbuilt environment in his discussion of the growth of Kennecott and its subsequent demise following the collapse of the world copper markets in the 1930s. Like many historians and geographers concerned with questions of nature and the complex politics and histories that complicate its meanings, I am influenced by Cronon's ideas and narrative strategies. But there is an addendum to the story of Kennecott's enterprise in the Northwest that might broaden the spatial and temporal scale of Cronon's concern with marginality, movement, and place.

After shutting down its Alaskan mine in 1938, Kennecott (now known as Rio Tinto) undertook an extensive geophysical survey program in northern British Columbia through its Canadian subsidiary Kennco Explorations (Western) Limited. Competing geologists discovered promising mineralization close to the confluence of the Stikine and Iskut Rivers in 1955. This prompted Kennco to develop its own Iskut–Stikine geochemical reconnaissance program.[68] The results convinced Kennco to partner with competing interests with stakes in the area, Cominco and Hudson Bay Mining, to form Stikine Copper in the early 1960s. Kennco explored on the property throughout the decade, diamond drilling from 235 holes and over eight hundred metres of underground development, and building an access road to the exploration camp.[69] Intermittent exploration for the next few decades confirmed the general character of the mineral deposit, but issues of access, topography, and the scale of necessary investment derailed any effort to build a working mine. NovaGold (at the time SpectrumGold) joined the partnership in 2003, with an option to buy Galore Creek, as the project was then called, outright from Stikine Copper. The dazzling success of NovaGold's exploration campaign convinced Teck Resources (formerly Teck Cominco) to join the partnership. An impact-benefit agreement was negotiated with Tahltan Central Council, the de facto resource leadership unit of the Tahltan. This whirlwind of ownership changes is routine in the mining industry, where promises are often as valuable as proven geophysical data. But Galore Creek has both promise and proof. It is thought to be one of the largest undeveloped copper and gold properties in the world, with almost seven billion pounds of copper, four million ounces of gold, and sixty-six million ounces of silver for good measure. "Measured and indicated" resource totals raise those figures by orders of magnitude.[70] As it stands, the operators of Galore Creek are waiting at the threshold of an immense industrial undertaking – for infrastructure such as access roads and power from the provincial grid, as well as final approval from government bodies tasked with assessing the feasibility of the project components and the social acceptability of its environmental footprint.

The Stikine is currently at the heart of the next great mining boom in the north. The search for and extraction of copper, gold, and coal will drive the resource politics of the region for the foreseeable future and will connect the Stikine to the rest of the world through commodity markets, consumption patterns, resource highways, and electrification schemes that give power to mineral extraction projects, making them not only possible but hugely profitable. By following the paths in and out of the Stikine – the roads, railways, and transmission lines – and showing how they connect to both the towns and mines of the Stikine as well as the minds of Stikine residents, I hope to shed light on the history and politics of resource use in northwest British Columbia. And by showing how these envisioned paths helped to move things through the Stikine, I hope to illuminate the role that the side effects of movement and failure have on the environment and its inhabitants.

1
Cassiar, Asbestos: How to Know a Place

Cassiar, 2010

There are no road signs to Cassiar anymore, but an affable gas-station attendant tells me to turn off just before Jade City, 140 kilometres north of Dease Lake on the Stewart-Cassiar Highway. It is late June, and the road into Cassiar is in reasonably good shape considering it has been left unmaintained since Cassiar Asbestos, the owner of the mine and town, declared bankruptcy and closed both in the middle of the winter of 1992. The road meanders along the bottom of an alpine valley, past the detritus of abandoned buildings and equipment-servicing operations, until it crests a hill, offering a first glimpse of the tailings pile. Nestled in the shadow of McDame Mountain, the site of the open pit asbestos mine that sustained the town for forty years, the tailings pile sits still and untouched, like a big grey-green toad. Although I expected it to be there, the pile of crushed ore is startling. Its outline is stark against the surrounding mountains. Even in this post-industrial nature, it seems utterly out of place.

I feel a strange disquiet as I approach the pile. Asbestos carries ominous connotations in contemporary society, as a health hazard and an environmental contaminant. But asbestos has a different history in northwest British Columbia. Cassiar was the primary employer, the main service centre, and the only permanent magnet for infrastructure development in the region between 1952 and 1992. Its history is intertwined with the social, economic, and environmental development of the Stikine in the latter half of the twentieth century. As I inch closer to the mountain of

FIGURE 1.1 Three Views of the Tailings Pile at Cassiar.

waste rock, I contemplate the particular constellation of asbestos, deindustrialization, abandonment, and corporate failure that characterized the lives of Cassiarites and determined the fate of the company in the late 1980s and early 1990s.

Cassiar is an emblematic unbuilt environment. It is not even a ghost town: most of the houses and buildings have been dismantled, sold at auction and taken to other sites in the region. The few buildings that still stand have been repurposed as storage facilities for prospectors or stand as neglected examples of antiquated mining and milling technology. The town that once existed is no longer there. Residents have long since been forced to leave. But the tailings pile is a stubborn testament to the environmental effects of the mine and the social impacts of the townsite. This chapter examines the environmental legacy of the mine and the economic circumstances that led to the abandonment of both asbestos production and permanent settlement in the Cassiar Mountains. I assess the development, dislocation, and re-emergence of "community" among residents and company employees. There is a peculiar convergence of environment and community memory in Cassiar. The townsite is dismantled and hidden behind a locked fence, but place-based connections internalized by former residents have persisted in spite of the physical dislocation that resulted from the mine shutdown. Cassiarites now live across the province and the country but they have come together online to unite under the banner of "Cassiar … Do You Remember?"[1] The recreated community, its virtual geography and deep associations with place, are important motivators for this chapter and its engagement with Cassiar's environment and the remarkable resolve of Cassiarities in the face of closure, economic failure, and displacement.

I stand outside my car, taking pictures and trying to rationalize the health hazards I imagine I am inflicting on my body by being near the pile without a mask. I am an asbestos neophyte, my general knowledge about asbestos running up against my research curiosity. A big, red pick-up pulls up beside me, the same truck that had already driven by a few times as I stood there. The truck window rolls down and the driver, a burly man of fifty with an awesome handlebar moustache, asks me what I am doing. His air of suspicion betrays the ongoing politics of mining in an area of uncertain Indigenous property rights and lingering environmental anxiety. I tell him, too quickly and too excitedly, that I am a researcher hoping to write about Cassiar and the mine and the tailings pile and asbestos. I share the few things I had read and found in the archives – about what Cassiar was, what it meant, and what was lost when the mine failed, and

FIGURE 1.2 Cassiar Tailings Pile.

I confess that although I had heard about the tailings pile I had never understood what it was until I had seen it ten minutes earlier. We talk for about fifteen minutes. He is a prospector with several gold and copper claims in the adjacent hills. Eventually, he decides to trust me. He had stopped, he says, to make sure I was not with the Sierra Club or some consultancy paid to quantify the value of untapped area resources.

He invites me to follow him to his office, located east of the tailings pile in the Storie Cabin, one of the first permanent buildings on the site. After introducing me to his employees and showing me some of the inner workings of his operation, he tells me to hop in his truck for a quick tour of the townsite and mill. If he does not take me, he says, I will never be able to see it. I will be safe, he assures me: he has been working there for ten years and does not know anybody who has gotten sick from exposure to asbestos. I leave my car at the cabin and begin the impromptu tour.

We drive back past the tailings pile, the base of it barely fifteen feet away, next to the fetid, roadside ditch. For the first time, I notice Troutline Creek, the primary water source and sometime industrial refuse dump, on the other side of the road. It looks clean and clear. We approach the gate through an empty parking lot. On the right, I can see the mottled red roof of the wet mill. On my companion's instruction, I unlock the gate and am told not to tell anyone who had brought me in. Directly in front of us is a blue corrugated tin building that was once one of the primary management offices.

We turn left past the hockey rink – or, at least, the rubble where the rink once stood. After the shut down, the arena was used by prospectors and miners to store equipment during the inactive winter season. In the winter of 2009, the unmaintained structure collapsed under the weight of snow, crushing the trucks, campers, drills, and shovels left in storage from the preceding summer. Now, the lot outside the former rink is a graveyard of twisted steel and derelict machinery.

South of the arena, the geometric street grid of the townsite is still imprinted on the landscape. We drive down the main thoroughfare of Connell Drive, still intact. The dozen or so side streets that once transected Connell have been dug up in an effort to return the valley bottom to some semblance of wilderness. The houses are all gone, either torn down or transported to proto-Cassiar subdivisions in Dease Lake, Iskut, or Watson Lake. Driveways, often surfaced with tailings material, are still visible, as are the edges of the flowerbeds, some of them with resilient perennials poking out from behind dessicated borders. We drive to the end of the road to look at the dump. It is on the other side of Troutline Creek, but the bridge has been taken down and the site is inaccessible from where we are. Troutline is small, shallow, and slow: I could wade across but my guide has a drilling appointment later and I do not want to impose further on his good will. The material history of settlement – house lots, waste deposits, flower gardens, minor league hockey games – exists in outline, but remains tantalizingly ephemeral.

We turn around and head back up the other major roadway, Malozemoff Avenue. The same landscape continues to my right but to the left is another example of the civic culture that the company promoted to entice workers and their families to a remote corner of British Columbia. A T-bar ski lift traces a route up the hillside, where workers skied during their free time. After several hundred metres, we arrive at the Catholic church, empty of pews and sermons but still solemn with the faint "material traces" of the iconography of worship. The windows, with the faded remnants of stained glass, remain unbroken though the brutal northern winters at an elevation of three thousand feet have chipped away at the building's white exterior paint. My guide's time seems less precious than I had initially

FIGURE 1.3 Cassiar Catholic Church.

thought; he keeps encouraging me to stop and take pictures of the places that interest me. He seems proud of what is left of Cassiar, claiming a kind of ownership over the place through his intermittent presence during the summer exploration season.

There is some evidence of recent activity in the distance. A late-70s-model gasoline transport truck, with the unmistakable Shell Oil clam logo, is parked in front of three long, brown apartment complexes still standing. These are the former "single quarters," where itinerant labourers were housed when they arrived, or where family men boarded until their families joined them or their houses were built. Now, the buildings are used to house contract diamond drillers, prospectors working their claims, and jade miners harvesting the waste debris of the asbestos mine. For $160 per night, the mostly male clientele get room and board and a warm shower, all powered by a generator. This is a welcome development in the new Cassiar, where the lack of infrastructure previously meant fairly Spartan conditions during the months of on-site work.

We continue east, away from town, towards the milling complex located beside the tailings pile. There are few buildings left intact, most prominently the wet mill and the dry rock storage. The mill buildings are not safe to approach and, having been sold at auction, they are now private property; my guide is reluctant to get too close lest someone report his further infraction of the unspoken code of conduct. On our way past the milling complex we pass under the aerial tramline, originally built in 1955 to haul broken ore from the mine to the mill. The baskets hang motionless in the air along the full three-mile line to the pit. Underneath the tramline is the makeshift processing facility of a private jade-mining operation. High-quality jade is everywhere in Cassiar. It was separated from the asbestos fibre by the company and used to pay for community amenities like the ski hill, the movie theatre, and the curling rink. Now it provides considerable income for a local prospector who has established his own jade monopoly in the region. Cassiar jade is sold around the world, much as Cassiar asbestos was sold globally in decades gone by.

This should have been the end of my time in Cassiar. I feel lucky to have seen the mill and the townsite, to have spent a few minutes inside the church and experience the eerie magnetism of the tailings pile. But my guide decides that I should see "the pit."

We turn north onto the mine road, the first bit of infrastructure built by the company. It runs adjacent to the tramline and was used to haul ore,

FIGURE 1.4 Cassiar Dry Storage and Wet Mill.

FIGURE 1.5 Cassiar Tramline

equipment, and personnel to and from the pit year-round, often all day and night. Frequent equipment upgrades designed to maximize production ensured that the road was in constant use and under heavy stress from eighty-tonne trucks in perpetual motion. Abandoned mining equipment, trucks, and camping gear are stockpiled along the route. The gear is in varying stages of disrepair. I am told by my guide that "Junkyard Larry" makes a decent living taking apart the equipment and selling it for parts or scrap in Smithers. After I learn this, I see Larry's imprint all over Cassiar. Piles of rubble take on new meaning as symbols of the new micro-economies that emerge in the wake of failure. Jade and junk are Cassiar's commodities now.

We make two stops on our way to the pit. About a kilometre and a half from the mill site we find the mining shovel that lived at the bottom of the pit for the entire life of the mine. It has been left in a field by the side of the road. The shovel from P and H Mining Equipment was the only shovel ever used by the company to remove the dislodged material into trucks and tramline cars, providing almost forty years of uninterrupted service, longer than the service of any individual mine worker. There is a certain romance to a piece of industrial equipment that outlasts its operators and its corporate owners. But the shovel will never be used again. It is now outdated; even if it were operational, it would cost too much to move it to another mining site. So it sits beside the road.

Further up the road is the crusher. Before being loaded onto trucks and tramline cars, ore would be broken into smaller pieces so that it could be better processed at the mill. The crusher improved production by no small measure. However, this machine, too, has been consigned to the same fate as the shovel – equal parts industrial waste, relic of a northwestern mining past, and big yellow warning of uncertain mining futures.

The road from townsite to pit is six kilometres long and climbs to an elevation of six thousand feet. As we climb McDame Mountain, we get an unimpeded view down the Cirque Valley, which would take us right to Atlin, British Columbia, site of a secondary gold rush in the early twentieth century. A barely navigable road snakes through the valley, a favourite hunting ground for those who work or who once worked out of Cassiar. Hunting and other outdoor activities were the primary leisure pursuit of Cassiarites. The possibilities of the surrounding landscape enticed many workers to Cassiar and were often cited in promotional material as an antidote to the blistering cold of winter. It is hard to imagine that hunting was any great solace to the men working the night shift in February at six thousand feet, in air so frigid that it made bolts snap right off the machinery and tramline cars crack in the cold.

My guide was full of anecdotes about Cassiar: people and places, practices and pastimes, gossip and scandal.

We do not make it to the pit. Someone (the jade miner was the primary suspect) had dug a large trench across the road about a kilometre before the new, reduced summit of McDame Mountain. My guide suggests the trench was probably for drainage, but, he speculated, it might just as easily have been intended to prevent other jade-seekers from scouring the pit and waste piles that had been pushed off the side of the mountain. So we back down the mountain, the road too narrow to turn the truck around. We are both disappointed, but seeing all of this is more than I had hoped for when I left Whitehorse for Cassiar that morning at 7 o'clock.

We come back through town and meet Junkyard Larry at his headquarters just outside the gate, behind the rink. We stop to chat with him as he loads gnarled steel onto his flatbed truck. It turns out there is organ-

Figure 1.6 Cassiar Crusher and Shovel #3.

izational method to the madness of rubble surrounding the arena. Larry admits that business prospects are very good this year. The bounty he has excavated from the rink will provide enough material for several summers' worth of runs to Smithers.

My guide has one more place to show me. We drive out of town, past the Storie Cabin and follow two ATVs down a secondary road that seems to lead nowhere. We emerge from the brush onto a long, straight tarmac. The airstrip, I am told, was used mostly for medical emergencies and for supplementary deliveries of goods and services. But for a few years, they were landing 737s with some regularity. My guide muses on the cultural differences he has witnessed at his various mining properties. If we were in Montana, he says, the tarmac would be filled with trucks piled high with barbeques, beer coolers, and hunting equipment. And we would hear gunshots, round after round, fired at targets in the adjacent bush. But in northwest British Columbia, the tarmac is empty. We turn the truck around and drive straight towards the tailings pile.

FIGURE 1.7 Driving on the Cassiar Airstrip.

Figure 1.8 Cassiar Asbestos Afterlives.

I leave my guide at his office. He goes to meet his contractors and I go for another, closer look at the tailings pile. I park my car, no longer feeling much apprehension about health, environment, or property. I jump across the ditch and I am on the pile. Paralleling my footprints are animal tracks, moose or bear by the indistinct size of them. The tailings pile has been the subject of several business proposals since the closure. There is still a mineable grade of asbestos in the pile, and new technologies could likely mill it economically. The ore is also filled with manganese, an element used as an alloy agent in steel and aluminum. My guide maintained that in the next ten years, someone would option the tailings pile and ship it wholesale to Asia for processing. Clearly, Cassiar's economic potential is not exhausted. This fact is obvious to the naked eye. Although you cannot see the manganese, the asbestos and jade are visible underfoot. Hairy, fist-sized emerald-coloured rocks are everywhere. I pick up dozens of samples before deciding on a few souvenirs.

It is time to leave Cassiar. I am going to be late for a dinner meeting in Dease Lake. But on the way out of town one last thing arrests me. I do not know how I missed it on the way in. Stashed on the side of the road, beside an old equipment garage, are several pallets of packaged asbestos. Split open and unused, they invoke the risks of industrial failure and the politics of mining in marginal areas. Each package bears the imprint of the Cassiar Asbestos Company.

Cassiar, 1952–92

Geology, Exploration, Discovery

In the summer of 1950, four prospecting partners staked seven claims on McDame Mountain in the Cassiar country, about one hundred kilometres south of Watson Lake. Hiram Nelson, Victor Sittler, and the two Kirk brothers were amateurs, poking around the region on weekends and their days off, but they were after a big find. They decided to investigate rumours of an asbestos outcropping – rumours that had been circulating ever since the men had come north to work on the Alaska Highway. The partners did not really "find" anything, but they did claim ownership over a commodity that had gained new value with its emerging industrial applications and its new accessibility, enabled by the highway that the men had helped to build.

The partners knew they lacked the capacity to bring the deposit to production, so they quickly sold their shares to Conwest Exploration. In May 1951, Conwest incorporated the Cassiar Asbestos Company, sent some fibres to Ottawa for sampling, submitted an additional thirty-three claims, and began to plan for more extensive diamond drilling in order to get a more complete rendering of the economic promise of the deposit. They determined that there was

> a large body of talus material containing asbestos fibre overlying serpentine rock. There is estimated to be 290,000 tons of asbestos ore available in the talus material, and exploration and development to date indicate that this is underlain by a large tonnage of serpentine rock carrying a good percentage of similar fibre. The results of testing from the talus material have indicated

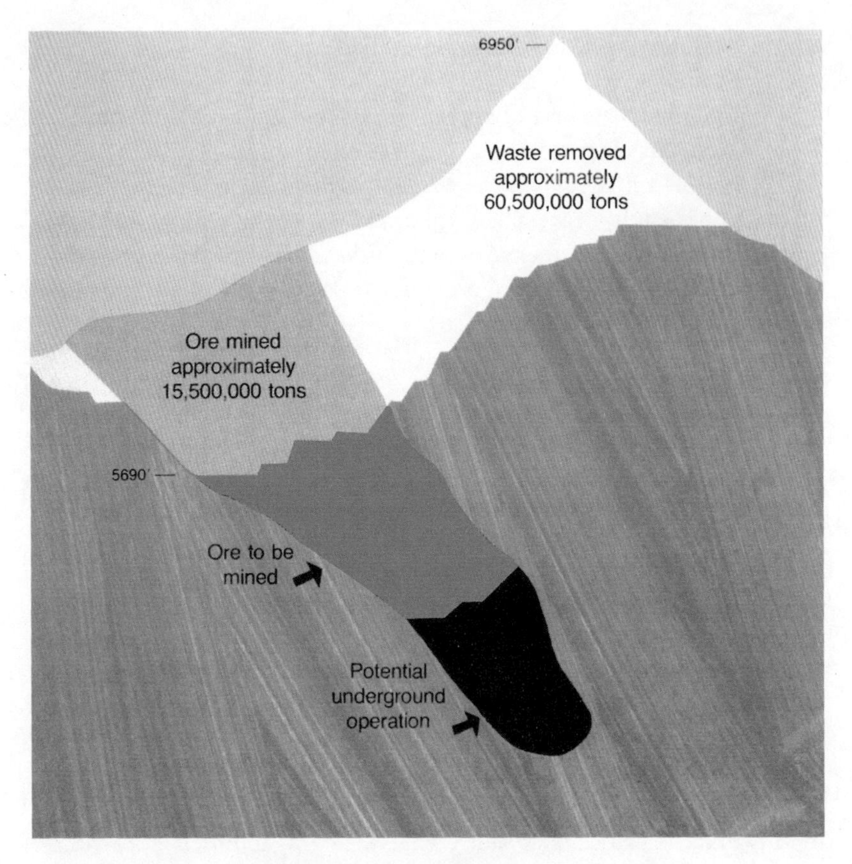

FIGURE 1.9 Cross-section Diagram of Cassiar Open Pit.
Source: George McLachlan (illustrator), Cassiar Employee Information Booklet, n.d, 13

> a high quality spinning fibre and a recovery of 8% fibre of this grade is indicated. It is expected that additional recovery of lower grade fibres will be practicable when operations are established.[2]

The grade and quantity of the fibre, plus the prospect of further reserves, motivated Conwest to build a mill capable of treating 250 tons of ore daily, with an initial annual production of approximately 6,000 tons of spinning-grade fibre by 1953. Additional drilling increased annually as deeper drilling exposed greater fibre reserves: 5,892,000 tons in 1952, and 7,232,000 tons in 1953.[3] Expenditures increased at a similar rate as mining and milling equipment was brought to the site and as transportation technologies were implemented. A gravity-fed steel chute was built from the lip of the pit, a power station and transmission lines were completed, an aerial tramway extending 14,600 feet from the pit to the mill was planned, and the adit was drilled deeper every summer, exposing greater reserves.[4] By 1962, after ten years of exponential increases in investment and capacity, the mill was processing 720,416 tons of ore annually and producing 2,357,623 tons of waste material.[5] Profits rose accordingly: from $4.7 million in 1954 to $115.1 million in 1971.[6]

This Cassiar mine development was, by most standards of industrial growth, a rapid and successful program that maximized potential and profit. But why did Cassiar Asbestos engineers and executives decide to proceed so quickly? What had they found that was so valuable and so immediately worthy of investment and endeavour? "Asbestos" is a general term used to describe a group of six hydrous magnesium-silicate materials renowned for their supple and soft yet strong fibres.[7] Asbestos had increased considerably in value after its use became widespread in building construction and in more specialized industrial applications, beginning in the mid-nineteenth century, because of its resistance to fire, corrosion, and friction. Asbestos was used in concrete, for instance, because it was heat- and sound-resistant. It acted as insulation, and the tensile strength of its long fibres amplified the structural integrity of buildings. The existence of asbestos on McDame Mountain had been known for eighty years, but the geographical marginality of the area had made economical production impossible.[8] The deposit suddenly became viable after the completion of the Alaska Highway. The peculiar qualities of the Cassiar deposit were also essential to its marketability. It was an iron-free chrysotile asbestos found in serpentinite rock. Non-ferrous asbestos was rare, at the time found only in South Africa and in small quantities in Arizona. It was identified as a strategic mineral by military interests, valued for its use in electric insulation, particularly on naval vessels.[9] The chrysotile form, or "white" asbestos, would gain increasing economic value due to its long fibres and rhetorical value as the "healthy" form of asbestos. The company claimed that chrysotile asbestos did not pose the same health hazards as "brown" or "blue" asbestos. To buttress the health claims, management was particularly proud of claiming that its product had been to the moon: apparently, the primary heat-reduction agent on the nose of the Apollo 11 spacecraft was asbestos mined at McDame Mountain.[10]

Community, Population, Labour

When Cassiar was operating at peak capacity, the population of the town hovered around twelve hundred people. In its early years, the population was mostly male, but gender and age demographics began to normalize as more families settled in the town.[11] The company built a community around the mine to house workers and their families. Home ownership was at the centre of this program. The company laid a townsite to the west of the mill and tailings pile. With limited experience, management personnel surveyed the house plots and streets that gave shape to Cassiar. Plots of land could be leased from the company, and building materials were subsidized for wageworkers. Building construction was the responsibility of the individual "homeowner." An unofficial exchange economy emerged, with plumbers, carpenters, and electricians all working on each other's houses. Single men lived together in dormitory accommodation and ate in the company cafeteria. Management also rented company-built houses. The town had a school and a hospital, a bank and a post office, a grocery and hardware store, and a liquor store, all of which attracted shoppers from across northwest British Columbia. An RCMP detachment dealt with whatever minimal disturbances occurred. There was a travel agency to facilitate the great exodus during summer holidays. Schmoo Daze, a weekend community celebration of snowshoe, ski and dog sled races, marksmanship, snow sculpture, curling bonspiels and, conspicuously, the outhouse races, eased the tedium of the dark, cold winter.

Company recruitment material emphasized activity, recreation, and community wellbeing while it tried to mitigate the effects of distance and isolation: "Because this is a relatively small community, somewhat isolated from the larger centres of amusement and leisure-time activity, no effort has been spared to develop a wide range of athletic, cultural and recreational pursuits. They are made available in hope that your spare time will be filled with interesting and rewarding pastimes."[12] In her fine popular history of Cassiar, Suzanne Leblanc details the material comforts available to Cassiarites, built by the company to foster a sense of community: the curling rink, the movie theatre, the community centre, a lounge and bar, a games room, a library, an auditorium, the ski hill, the hockey rink, a soccer field and two churches, Anglican and Catholic.[13] A club room held meetings for the Lions Club, the toastmasters, boy scouts, girl guides, majorettes, the dart club, the P.T.A., and the handicraft club, as well as the gun club, badminton club, and duplicate bridge club. For many workers, the main enticement was unencumbered access to prime hunting and fishing territories and other outdoor recreation possibilities. Others drank and gambled. Cassiar was similar to most remote northern resource communities, except with more of the leisure trappings of modernity readily available. Townsite amenities were funded by the company with profits from the sale of jade, found in association with asbestos in the ore of McDame Mountain. Jade thus helped to mitigate some of the climatic and environmental disquiet of daily life in Cassiar, where winter months would pass without sun and the temperature would dip dangerously low. After a time, the sale of jade became so profitable that the company was compelled to include it in its general income reports for the benefit of shareholders.

The mill and mine were staffed by men from all over the world. An early promotional piece reported that the company needed between 50 and 150 employees "composed chiefly of Canadians, including Indians from Telegraph Creek, Lower Post and Whitehorse, but also new Canadians, formerly displaced persons of various national origins."[14] The company did not necessarily solicit this ethnic diversity, but it became a marker of its workforce. In 1961, there were men from twenty-seven countries living and working at Cassiar.[15] Turnover was high, particularly in the early years, when much of the leisure and wellbeing infrastructure was still being developed. The climate proved too much for some, others were undone by the isolation, while many were itinerant workers simply seeking a high-paying job before moving on to the next enterprise. Turnover was above 100 percent annually during the 1960s – 125 percent in 1964, 175 percent in 1965. Bill Zemenchik, mine manager for almost the entire duration of operations at Cassiar, estimated that 250,000 people worked for Cassiar in various capacities.[16] Skilled workers were the hardest to recruit and often the first to take their skills elsewhere: Cassiar Asbestos offered higher wages and apprenticeship programs in an attempt to keep these workers. This was a business decision for Cassiar: high turnover translated to higher operating costs and higher industrial accident rates, often resulting from inexperience.

From a company standpoint, there were considerable labour difficulties on top of the problems associated with skill, numbers, and wages. The Cassiar Mine Mill and Allied Workers' Union, Local 927, was formed in early 1954. Mill and mine workers later joined the United Steel Workers, Local 6536. Strikes became more common in the 1970s as unionization spread from the mine and mill workers to other areas. In 1972, a strike over wages shut down production for three weeks. In 1975, another strike, largely over health concerns, shut down production again. In addition, outside strikes continually disrupted shipment and production. Labour-management tensions continued into the 1980s and contributed to the ultimate failure of the company.

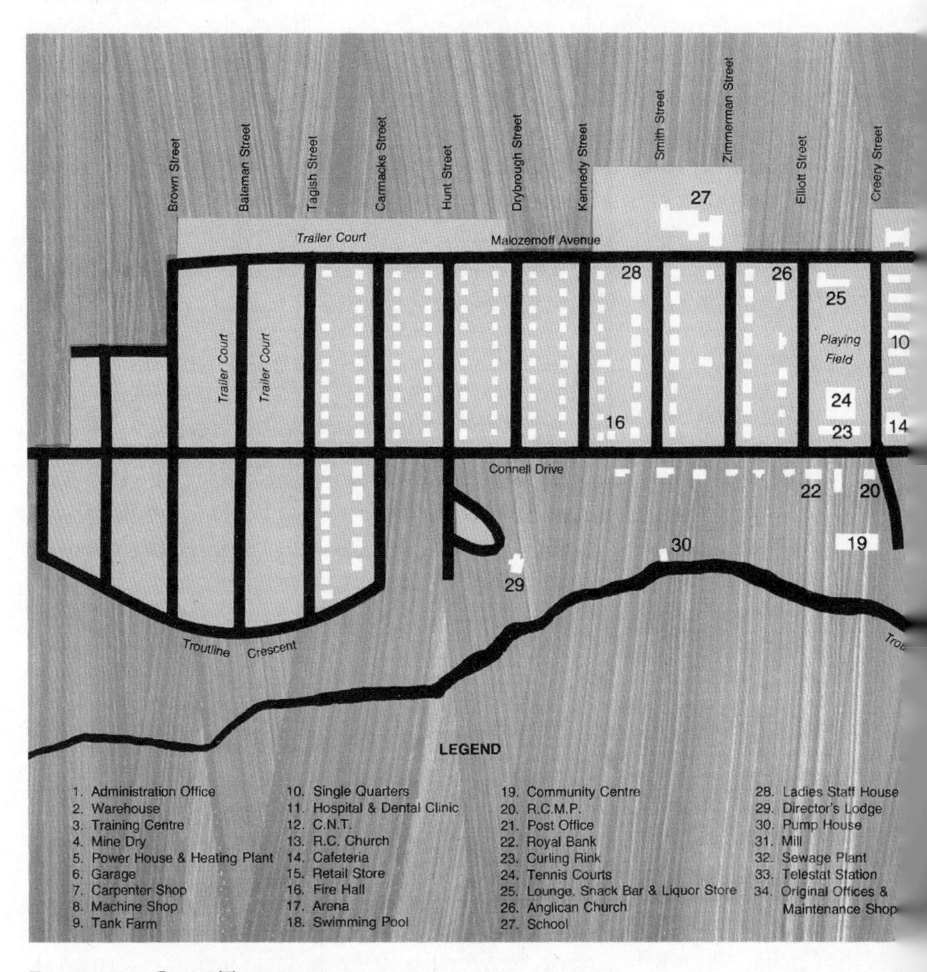

FIGURE 1.10 Cassiar Townsite.
Source: George McLachlan (illustrator), Cassiar Employee Information Booklet, n.d, 6–7.

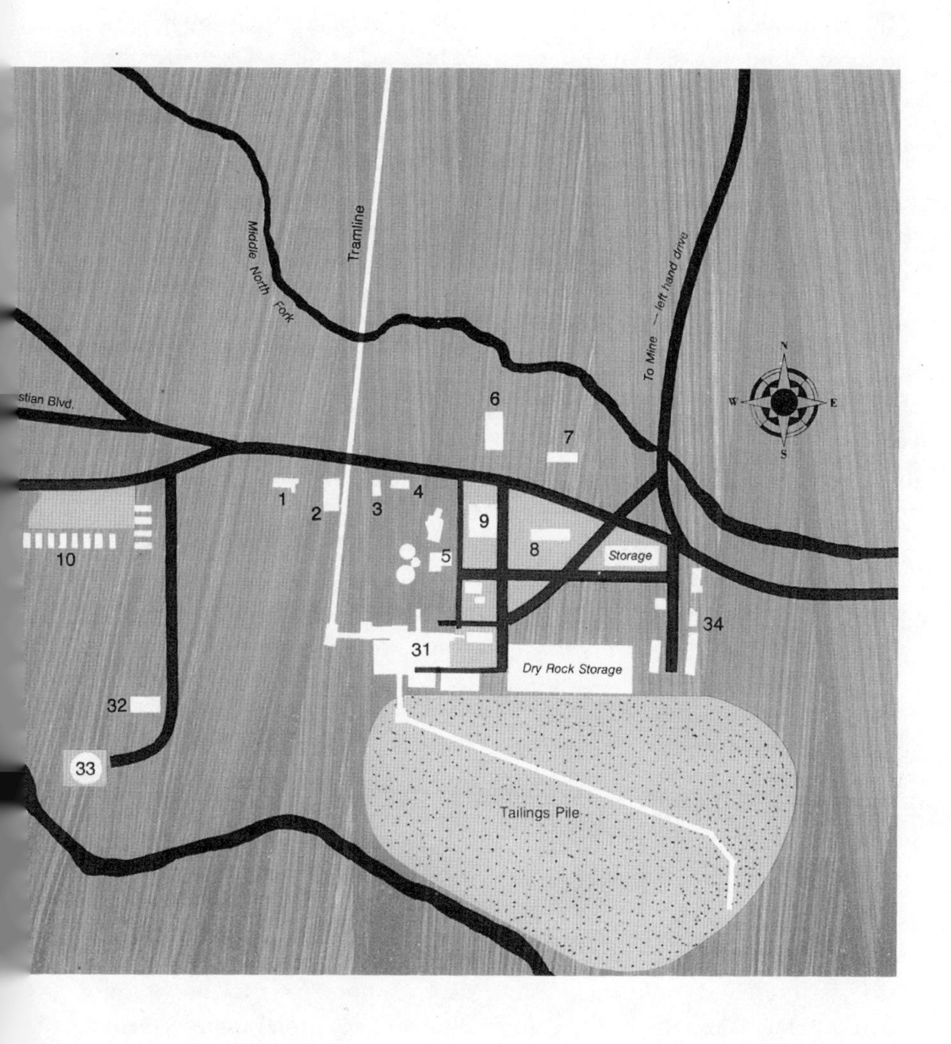

Middle North Fork
Tramline
To Mine — left hand drive
stian Blvd.
6
7
1
2
3
4
9
8
5
10
Storage
34
31
Dry Rock Storage
32
33
Tailings Pile
N
S
E
W

Transportation

From an infrastructure point of view, Cassiar was in the middle of nowhere. The economical and efficient movement of industrial materials and household goods into Cassiar and the export of finished product was a constant concern of management. A road into camp had to built at the outset, but it was not carved out of the wilderness. A Sacramento-based mining interest, Moccasin Mines, had a small placer gold operation in the vicinity in the late 1940s and had built a very rough road south from Watson Lake to bring machinery and supplies to its camp. There was no surveyor available so the company had simply chosen the route on its own.[17]

Immediately after staking their claim, Nelson, Sittler, and the Kirk brothers secured a thousand-dollar grant from the provincial government to improve and extend to Jade City the makeshift Cassiar Trail (as the Mocassin Mines road was known), and to build a branch road into the Cassiar townsite. This project was completed within two years. The company also used grant money to improve the road to the pit, originally built by Moccasin as a bulldozer road up McDame Mountain in 1948. On these roads, the bulk of the extracted ore would move for the next forty years.

The switchback road to the open pit was almost twice as long as the six kilometres that separated the mine from the mill. By the summer of 1952, the company had acquired twelve seven-ton trucks that were employed exclusively on the mine road, bringing ore to the mill for processing.[18] The size and capacity of these trucks were upgraded over the years of operation. By the mid-1970s, a fleet of eighty-ton trucks was in perpetual motion on the mine road, servicing the mill twenty-four hours a day, seven days a week. Aside from gasoline, tires were the major cost incurred by the transportation division. Tires were vulnerable to rock and sharp mine debris, particularly in the deep cold of the Cassiar winter. The road could be hazardous in winter, as drivers made the seven-hundred-metre decline between pit and mill.

Cassiar received considerable infrastructure support from the provincial government, especially in the construction and continual improvement of the Stewart-Cassiar Highway. The original shipments of asbestos went north via the improved Moccasin-Cassiar Trail to the Alaska Highway and on to Whitehorse, almost six hundred kilometres away, where they would be loaded onto the Yukon and White Pass Railroad bound for Skagway. From there, Cassiar asbestos was shipped to various Pacific markets or to Vancouver, where it was packaged again for onward movement. A small amount of product travelled east along the Alaska Highway, bound for Edmonton and markets in Canada and the United States. Construction of the Stewart-Cassiar Highway, explicitly for the purpose of providing Cassiar Asbestos with a direct link to a Canadian deep seaport, began in 1958 but was not completed until 1972 at a cost of $30 million. This reduced the travel distance between Cassiar and Vancouver by 650 kilometres, but inconsistent road quality, periodic washouts, and unreliable maintenance meant that Cassiar still looked north instead of south to Stewart.[19] In spite of its consistent lobbying for the road, the company did not use it until 1978, when it became a more cost-effective transshipment corridor than the Alaska Highway route.[20]

The Alaska Highway route was undone by its length, load-weight limits, a political dispute over territorial jurisdiction, and dust. The Stewart-Cassiar route was shorter, but until the early 1970s, the federal government (which collected duties on the Alaska Highway) offered Cassiar considerable concessions at the scales that monitored industrial traffic. All of the twenty-two trucks that Cassiar employed to make the run between the mine and Whitehorse operated over the legal weight limit, with the full knowledge of federal authorities. Cassiar paid fees for oversize trucking on the BC portion of the route, but took advantage of a "policy of unofficial "tolerance" without overload permits."[21] It certainly helped that there was no scale on the Yukon portion of the route until December of 1970.[22] When supervisory responsibility over the Alaska Highway was devolved to the territorial government, there was tangible anxiety at Cassiar that load limits would be enforced and the advantages of the northern route would be diminished. Still, this change would take the better part of a decade, as well as improvements to the port at Stewart, before asbestos traffic was re-routed from Whitehorse.

Dust also played a role in the decision to use Stewart as the main terminus for shipment to Vancouver. On the northern route, "Yukon Dust" seeped through "the smallest crack or opening contaminating everything."[23] The product got dirty on the trip to Whitehorse, masking the "whiteness" that made Cassiar asbestos so valuable as a commodity. The solution was simple: it was less dusty on the southerly route, and "palletized shrink wrap" packaging technology further protected the containerized asbestos fibre and also reduced losses during transportation.[24] In a further effort to accommodate Cassiar's shipment needs at Stewart, the provincial department of highways allocated $40 million to upgrade and widen the Stewart-Cassiar Highway. The province also offered truckload-weight concessions that Yukon could not match. Concurrently, BC Rail's Dease Lake Extension was under construction, with the assumption that Cassiar Asbestos would be a major customer. Perhaps because Cassiar was in the extraction vanguard, it seems clear that the company was in a favourable transport position, with governments and various government departments competing to provide the company with the best competitive advantage.

Environment

"The environment" was not an important administrative concept in the early years of operation at Cassiar. Processing and production overrode concerns about environmental pollution. There was some awareness of the dangers of ambient dust, and measures were taken to collect and disperse dust during the milling operation.[25] But this was a localized response to a regional problem. Former employees remember a simple method of dating snowfalls: a shaft dug into the snow exposed lines of green, where mill dust had fallen on top of each snowfall. A photograph from the 1960s shows a dust cloud hanging over the town on an otherwise clear day. Prevailing winds were westerly, taking most of the dust away from town, but much remained, hovering above the townsite and touching down when winds subsided. Anybody living in the cabins and shacks east of town and the mill weathered the brunt of the hovering dust.

FIGURE 1.11 Cassiar in the 1960s. The view here is from the top of the ski hill.
Source: Northern BC Archives, Cassiar Asbestos Mining Corporation fonds, Accession #2000.1.1.3.13.4.

By the mid-1970s, the environmental effects of open pit mining and asbestos production had become central to the operational organization of the mine and the townsite, as perceptions of asbestos shifted from mineral to hazard. Policy modifications were as much a result of outside pressure as they were of broader, social-ecological concerns. The US Environmental Protection Agency (EPA) led the charge with a forceful critique of asbestos and its health risks, and Cassiar Asbestos' 1974 annual report highlighted a developing environmental program. The company created a new department of environmental affairs that was concerned generally with waste removal systems, reducing airborne fibre counts, dust prevention, and improving standards of cleanliness.

These innovations were instigated by the looming possibility of government-enforced decreases in allowable asbestos dust standards. When the BC Mine Inspection Branch imposed a more stringent guideline of five fibres per cubic centimetre (as opposed to five million fibres per cubic foot), the company was forced to shut down operations for three weeks to meet the new standards. It was intimated that the standard might be further reduced to two fibres per cubic centimetre: the company complained that this limit was impractical and that the technology required to meet the target was either not available or too expensive. Yet by 1981, the company had met the prospective standard and was proud of the expenditure and achievement: "over $18 million has been spent at the Cassiar asbestos mine on environmental facilities and equipment. These capital expenditures for modifications and improvements within the mill complex have resulted in the reduction of overall average fibre concentrations to less than two fibres per cubic centimeter, in compliance with BC government standards."[26]

A two-tiered environmental committee, composed of community, union, and management representatives, was formed in January 1976 to supplement the environmental affairs department and to consult employees in environment and health matters. The committee published newsletters designed to "get to the heart of the matter" without "complicated and somewhat mysterious" medical jargon and the fear-mongering of the media and the EPA.[27] In the newsletter, Garry Doran, Cassiar's newly hired environmental engineer, admitted that there were potential environmental side effects of asbestos production, but these were mitigated by the type of asbestos at Cassiar, the production techniques used, and new technologies mobilized to alleviate the most potent dangers. The new program included:

"a) a continuous monitoring program, b) extensive medical examinations, c) new and improved equipment enclosures, d) the use of new and improved dust lifts, e) protective clothing, f) confining asbestos to the working environment, g) personnel dedusting booths, h) a new and improved vacuum cleaning system, i) mobile equipment cleaning, j) conditioning of wastes to reduce emissions, k) education of the employee[s], l) an anti-smoking campaign, m) improved mill circuitry."[28]

The environmental committee itemized the company response to environmental dangers while downplaying the actual effects on area ecosystems. The response was publicized, but the dangers remained ambiguous and undetermined. Cassiar would also fight for the remainder of its days the "erroneous" decisions of the EPA to warn against and ultimately ban the use of asbestos in the US. By the mid-1980s, Cassiar could point to the Ontario Royal Commission on Asbestos and various industry-affiliated organizations to claim that chrysotile asbestos posed no discernible environmental or public health risks.[29]

Cassiar was forced to enumerate specific areas of environmental concern in a submission to the Pollution Control Board of British Columbia during a hearing on the reduction of fibre emission objectives in January 1978. This was a fairly comprehensive document that catalogued discharges into water and atmosphere and discharges onto land for the mine site, plant site, and townsite. In general, Cassiar gave itself a passing grade. The main site of environmental contamination was the tailings pile, which covered forty-five acres of the valley floor and contained twelve million tonnes of material. It was acknowledged that there was some danger associated with the pile, but the document emphasized the money spent on mitigation programs. The submission claimed that operations at Cassiar benefited from increased environmental awareness because "environmental and process improvement tend to go hand in hand, the process improvement allowing generation of the capital required."[30] This tactic allowed Cassiar to issue new warnings about the dangers of over-regulation:

> Increasingly large amounts of capital have been expended on environmentally related projects at Cassiar operations. In fact, the capital requirements for these and future projects are becoming so large that Cassiar's known ore reserves may not be able to support all of them. This potential problem is of vital concern to all persons or agencies directly or indirectly deriving benefit from this primary resource industry.[31]

Cassiar supported environmental initiatives, but only to the point that did not interfere with production.

The submission also highlighted the company's relationship with Troutline Creek, which was a flashpoint of environmental concern. Troutline was at risk because of its proximity to the tailings pile; it needed to be protected because it was the main source of drinking water for the community. Engineers employed an industrial solution to the problem: "Between 1973 and 1976, diversion work was carried out to relocate Troutline Creek to a point further south of the Plantsite. This eliminated possible water contamination from the tailings. A restraining berm was installed to impound runoff during springtimes."[32] By bulldozing a new creek bed, Cassiar claimed to ensure the safety of the water. Troutline Creek was subject to company testing and was always considered safe.

Independent testing at another Cassiar-owned asbestos mine in the Yukon revealed some health concerns. Residents of Clinton Creek, where a mine operated from 1967 to 1978, asked the federal department of environment to investigate the water they were drinking. That study concluded that there were "significant adverse effects on the invertebrate communities and fish populations as a result of substrate alteration and the presence of asbestos fibres in the vicinity of the mine ... the water chemistry of the area near the tailings and waste rock pile has been affected as indicated by elevated levels of calcium, magnesium, iron manganese, potassium, turbidity and hardness."[33] Any similarities with the situation at Cassiar are a matter of conjecture, but it is clear that water-quality standards were a contested matter open to some interpretation. The tailings pile and other industrial emissions became part of area ecosystems and interacted with human and animal bodies in ways generally not policed by the company or environmental regulators.

By the late 1980s, Cassiar had begun to develop more comprehensive environmental policies. Pollution policy would no longer be considered in isolation but would permeate all levels of corporate planning. The company developed a four-pronged strategy that focused on air, water, and noise pollution, and the abatement of solid waste, but it would also focus resources on communications, organization and training industrial health, customer protection, and defining problems and monitoring systems. Communication was understood to be "at the heart of any successful system." Yet the bottom line was still clear, even if the corporate language had changed: "Abatement planning must be a balance of the equities between corporate economics and desired environmental quality."[34]

The tailings pile was the ever-present green-grey elephant in the room. Dust and water contamination were certainly on the corporate radar of Cassiar Asbestos. But the real environmental legacy of the mining operation and the tailings pile remains to be seen. And it is an ongoing story. Both the pit and the pile were left unremediated. Groundwater seeps up into the pit, rainwater and melt water create a pond every spring. Animals walk across the pile, fish swim in Troutline Creek across the road from the pile, and prospectors drink its water.

Health

The question of asbestos-related health dangers is intimately tied to the environmental questions that became so threatening to the asbestos industry. For Cassiar Asbestos, health and environment went hand in hand: interest in the health of workers and customers paralleled the new concern with environmental issues. Environmental innovations were more closely associated with limiting health risks than with mitigating ecological damage. Concern over employee health followed roughly the same timeline as environmental planning. In early years, despite a general awareness of the health risks associated with asbestos production, the company did not have a concerted health policy. Health as a topic seems to have been addressed under the title of safety, which was essentially the minimization of risk: "All new employees are required to attend a Safety Indoctrination meeting. This requirement must be met as very important information on Company policies, requirements and practices, fire control, safety and legal aspects are explained."[35]

The company instituted health policies as the specific health risks of long-term asbestos exposure – lung-related cancers, asbestosis, mesothelioma – became better known. The modernization program of the mid-1970s included funds for the radical reduction of ambient dust in the mill. The mill was enclosed to become a self-contained unit, a vacuum system was installed to facilitate cleaning, air blowers maintained circulation, and mill air pressure was redesigned to reduce the escape of fibres from inside the mill to outside. Much like the company's environmental improvements, these health measures were seen as "non-productive" but necessary in light of impending legislative adjustments and the possibility of lawsuits.

By the late 1980s, as the costs from lawsuits reached upwards of $10,000,000, Cassiar was continually forced to defend its health record.[36] The company mobilized examples from science and history – particularly highlighting the uniquely safe properties of chrysotile asbestos – to make its case. The company told the Workers' Compensation Board of BC in January 1989 that

> A review of the relevant scientific evidence reveals that adverse health effects have not been observed from exposure to only chrysotile fibre in the relevant industrial settings at levels near, but still substantially higher than, the current limit of 2 f/ml [2 cm^3]. Thus, the present limit of 2 f/ml is clearly sufficient to protect workers from asbestosis ... Moreover, there is no basis in the scientific literature for concluding that chrysotile has ever caused mesothelioma at such low levels as 2 f/ml ... Finally, in studies of well-controlled factories using only chrysotile, no excess of any kind of cancer has been observed at levels such as 5 f/ml.[37]

Anecdotal evidence challenges such assertions. There are several reported cases (or at least cases referenced by former employees) of asbestos-related diseases among workers at Cassiar, and much suggestion that other cases must exist, particularly among the itinerant Indigenous community located downwind and downstream from the tailings pile.[38] Company health records are sealed and liability is limited and difficult to prove considering the company was dissolved two decades ago.

McDame

The open pit mine always had a finite life. Engineers projected a forty-year lifespan when the Cassiar mine opened, and that projection was upheld as new ore discoveries over the years dovetailed with advances in production technology. In the late 1970s, during a time of uncertainty in the Canadian mining sector, engineers stumbled upon a second asbestos deposit while conducting seismic studies on the stability of the pit walls. After years of careful appraisal, company directors took a decision intended to extend the life of Cassiar for at least another ten years. The construction of an underground mine adjacent to the open pit was supposed to save the town and the company, but repeated engineering failures, financial difficulties, and market downfalls signaled the end for both.

The second major asbestos mineralization on McDame Mountain was found by accident in 1978. Cassiar had been having problems dewatering the open pit. The excess water was disrupting the mining and benching sequence as they extended the pit deeper. Company engineers concluded that the appropriate solution was to drill a new adit on the side of the mountain and, from this adit, drill holes into the pit to drain the bottom.

While the adit was being developed, the company decided to do some exploratory diamond drilling. They found the extension deposit and confirmed it the following year with a more ambitious drilling program. The news was greeted with considerable enthusiasm in Cassiar, as the newly discovered deposit was thought to ensure the immediate future of the community.

With this information, Brinco took over majority ownership of Cassiar in a corporate merger in August 1980. Brinco invested heavily in the development of the new ore discovery, known as the McDame vein. The drilling program was elaborated in 1980–81, and this was followed by an extensive air and ground magnetic survey. A series of pre-feasibility studies dampened optimism by raising doubts about the fibre grade of the deposit and the integral strength of the surrounding material.[39] Further testing revealed a deposit 1.8 kilometres in length and 1 kilometre wide, containing some thirty-two million tonnes of ore at a grade slightly below the asbestos from the open pit.[40] Al Burgoyne, the Brinco geologist leading the McDame exploration, stated in May of 1986: "A production decision, utilizing block caving mining methods, will be contingent on continuing favourable exploration results, mine operating costs, financing and markets."[41] By the end of the year, Brinco was boasting to shareholders that McDame was "technically feasible."[42]

A feasibility study published in December 1987 determined that there was sufficient ore to proceed and to ensure the life of Cassiar for a decade after the closure of the pit. The decision to proceed was now in the hands of Princeton Mines, a mining conglomerate under the corporate control of Toronto investment counseling firm Hamblin Watsa, which had bought a controlling share in Cassiar in February 1987. Under the leadership of newly recruited president and CEO James O'Rourke, management recognized the considerable risk and investment involved in the development of an underground mine. Expertise was lacking. Few of the miners had worked underground and no one in management had experience in building and operating underground. Unexpectedly, given that it was an asbestos initiative, financing proved relatively easy. In March 1988, provincial Minister of Mines, Energy and Petroleum Resources Jack Davis announced that McDame would receive a government loan of $25 million. The Bank of Montreal guaranteed loans of $12 million. Cassiar was left to find the rest of the $43 million in start-up costs.[43]

Construction began in June. After a few complications with the hired contractor, Cassiar staff were brought in to finish the job. McDame was completed on time in November 1990, although it was $20 million over budget. But production problems arose from the outset. It was difficult to break the ore into sufficiently small blocks to facilitate its removal from the mine. There was also considerable tension over the negotiation of a new union agreement. Having agreed to suspend pay increases over the last three contracts, the union was unwilling to make further concessions now that the underground mine was operational. The company argued that it could not afford a wage increase. A two-week strike was settled in March, but the work stoppage exacerbated a water inflow problem that had plagued the mine in the opening months of operation. The seepage problem had been contained by pumping, but the strike had halted regular operations and the mine was flooded when work resumed. In addition, asbestos continued to take a public relations beating. The EPA decided to ban most uses of asbestos for seven years, though the company claimed this would have little effect on sales, as the US was no longer a major market for its product. But the mine was also operating over budget, and the fibre grade was lower than what had been predicted. Yet the mine met its contractual obligations, and the company felt confident enough in the future profitability of the mine to host a 600-person mine opening celebration in August 1991.[44]

There have been several other attempts to mine out of Cassiar, though none have sought to reopen McDame. In the summer of 1994, a partnership called BC Chrysotile Corporation began processing the tailings pile for asbestos fibre that had been left over from open pit operations. The waste rock, sixteen million tonnes of crushed material, was believed to contain enough asbestos to be profitable. As part of the mining agreement, BC Chrysotile agreed to clean up contaminated soils in the townsite and to begin the process of turning the abandoned and stripped townsite "back to nature," saving the province an estimated $5–7 million in remediation costs. Roads were dug up, debris and garbage were organized and removed. Forty years of community development were consigned to existence as heaps of rubble, obscuring the most noticeable traces of settlement. In the late 1990s, the same company under a different name, Cassiar Mines and Metals, began a similar operation. Using the same milling facilities, Cassiar Mines and Metals was shipping product by the spring of 2000. With the planned construction of a new wet mill facility, Cassiar Mines and Metals planned to boost fibre production to fifty thousand tonnes annually, almost half of Cassiar Asbestos' peak production levels. The project was ultimately unsuccessful, but the "rehabilitation" of the valley floor transformed the landscape at the Cassiar townsite.[45] In 2003, Cassiar Resources, as the company was then known, sold its operation to Cassiar Jade Contracting, which currently mines, maintains, and polices the property. In Cassiar today, "green is the new gold" – definitely catchier than "green is the new asbestos" – and even the Discovery Channel has developed a case of "jade fever" with its reality show of the same name, which follows the Bunce family's jade mining business.[46]

Some speculate that the tailings pile at the Cassiar site will be optioned in the next decade. One local prospector told me that Chinese and Korean companies had been analyzing its contents with an eye to moving the entire pile to Asia in order for it to be processed there. A significant aspect of this new business rationale is the presence in the tailings pile of large quantities of manganese, an element with important metal alloy uses.

Reorganization, Failure, Closure

The development of the McDame mine was tied to the continual corporate re-organization of Cassiar and to an evolving corporate strategy meant to address shifting global demand for asbestos products.[47] Both Brinco and Princeton Mines sought to diversify corporate holdings to increase financial assets and liquidity and to distance Cassiar from the social stigma of asbestos. To this end, Cassiar acquired San Antonio Gold Mine in Manitoba; Newmont Mining, which operated the Similco Mine near Princeton, BC; and a 50 percent stake in Western Canadian Mining and the Lac Knife Graphite Project in Quebec. These investments resulted in Cassiar's considerable investment leverage, reduced debt, and increased cash flow, and directly facilitated its decision to proceed with the McDame project. When Princeton Mines took over in 1987, it completely reoriented marketing patterns. *BC Business* magazine compared Cassiar to cigarette companies. Like cigarette companies, "[Cassiar's] marketing strategy has shifted to where the action is" – away from North America and Australia, "the areas most sensitive to environmental and health concerns," and towards Asia and Eastern Europe, with additional target markets in South America and the Middle East.[48] James O'Rourke rejected the suggestion that this was a cynical corporate strategy directed at the Global South: "The reason that sales are climbing in countries like Indonesia and India is not only that they have highly sophisticated and growing construction industries, but because a favoured building product in such a climate is asbestos cement."[49] According to Cassiar, re-structuring and diversification was driven not by a desire to get out of asbestos but by opportunities presented in other arenas.

This is the corporate context of the McDame failure. Cassiar was able to secure financing for McDame because of its diversification program;

from a corporate standpoint, however, the business was undermined by the loans and differing political and business opinions about the economic viability of McDame. As collateral for the $25 million loan received in March 1988, Cassiar put up its wharf facility in North Vancouver, with the understanding that this facility would be returned to Cassiar once construction of the underground mine was complete and once pre-determined production goals were met. The company believed it had fulfilled these conditions. In order to meet further production targets and turn a profit on the McDame ore, Cassiar needed to hire fifty additional workers. It required $10 million to do so, and planned to raise half this amount on the market with an equity issue and half by selling the wharf to the federal government. There was, however, disagreement about Cassiar's progress. The provincial government claimed Cassiar had done little to repay the loan, and would not release the company from its collateral obligations. The company, faced with a lack of capital and labour, decided nevertheless to proceed with the McDame mine. The Bank of Montreal also raised concerns about its loan, sending the company a letter reminding it that the loan could be recalled at any time. O'Rourke met with government officials in a second attempt to get the wharf released so that the company could raise operating capital. An election campaign was underway and the Social Credit government, keenly aware of its likely electoral defeat, asked for an outside report on the viability of McDame. It commissioned Coopers and Lybrand and released a document that was interpreted very differently by the mine owners and the province. Cassiar contended that the report affirmed its plans and highlighted the market value of the ore. The province, now governed by the NDP, acknowledged the quality and size of the ore reserve but emphasized Coopers and Lybrand's skepticism about the economic viability of the mine. The government and the mine reached a standstill over the release of the wharf and, without the needed capital, Cassiar began to lay off workers and reduce its production.

The company filed for creditor protection in early October 1991. This gave the company three months to develop a plan to meet its credit obligations and restart full production. The "reorganization plan" claimed that Princeton Mines had no resources to continue operation of the mine and would be forced to shut down if it did not receive an additional $17 million loan from the provincial government to be drawn over the first eight months of 1992. Cassiar declared that the "consequences of not proceeding with the plan" would force the company into bankruptcy, resulting in "[a] loss of approximately 500 direct jobs, and the consequential loss of many other jobs, curtailment of services of the townsite including heat, light and power and the destruction of the economic base of Cassiar."[50] The company outlined steps to cut operating costs: customers would need to accept product cost increases, workers would be temporarily laid off, planned wage increases would be delayed, and creditors would have to agree to significant concessions. But the plan was rejected, and the government refused to finance the full measure of a reorganization in which shareholders would not share any of the economic risk. The NDP government under Mike Harcourt was trying to govern in a fiscally responsible manner after the previous Social Credit government had left behind the largest deficit in BC history. There was little money to spare for corporate bailouts.

Without funds from government sources, and unable to raise money privately, Cassiar declared bankruptcy in late January 1992. A community coalition, McDame Mine Limited, was formed in order to attract prospective buyers. The coalition garnered some interest, particularly from Black Swan Gold Mines of Vancouver, but ultimately it was a fruitless pursuit. Residents were given until July 30 to leave Cassiar or face eviction from their homes. Maynards, an auction house in Vancouver, was hired to sell everything. The new school building in Cassiar was disassembled and shipped, piece by piece, to Hudson's Hope in the Peace River valley. A developer from Sechelt bought lighting, telephone, electrical, and fire-prevention infrastructure for a subdivision on the Sunshine Coast. Mining companies from all over the world arrived to bid on the mill buildings, equipment, and trucks. The auction raised only $6 million, a figure far below that of Cassiar's debts. Most of what was not sold was bulldozed.

Cassiar's Future: Archives and Memory

In the summer of 1993, as part of the liquidation and dispersal of company assets, the University of Northern British Columbia (UNBC) agreed to house the records of the Cassiar mine and townsite. Files were placed indiscriminately in boxes, taken from offices and from the on-site storage warehouse, and loaded into trucks to be taken to Prince George. In all, a trove of several thousand bankers' boxes of archival material was delivered to the university. There was no logical organization to the material. Anything deemed worthy of preservation was simply removed from office shelves and filing cabinets and deposited in the nearest box. Organization and cataloguing of the material was left to the archivists at UNBC. The boxes waited until 2000, when archivist Michael Taft, seeking to demystify the "archeological remains of Cassiar," hired the first of many summer co-op students to begin sifting, identifying, and cataloguing Cassiar's disordered past.[51]

Engineering and annual reports, employment and health records, environmental assessments, and promotional material, production statistics and geotechnical data, maps and photographs: this is a massive collection chronicling the everyday operations of the mine and everyday life in Cassiar itself. It is an invaluable repository for someone wanting to write a comprehensive history of the company and the town. But until very recently, such a study was circumscribed by legal, administrative, and technical limitations on the archive. The Cassiar Asbestos Corporation archive was constrained by issues of access, accountability, and the terms and organization imposed by the UNBC archive and the lawyers hired to protect company interests. These constraints will seem old hat to historians used to navigating the controls imposed on corporate archives,[52] but they nonetheless set substantial obstacles in the way of research progress.

Material in the Cassiar archives was difficult to access. Early on, UNBC archivists made the decision to use a database to catalogue the collection. The contracted computer consultant, along with the managing archivist, chose the original version of Microsoft Access. It was an unwieldy program that circumscribed the organization of the materials into conventional categories – name, date, title, category, and so on – without offering the researcher the means to cross-reference. There was no detailed finding aid, so researchers, with the help of the archivist, had to methodically search individual terms or names to find relevant materials. The task of populating the database with Cassiar materials was turned over to a succession of summer and co-op students, who were not trained archivists or well versed in the corporate structure and management of the Cassiar mine. Some of the cataloguing would have been straightforward: the identification of report author or title, for example, or the date of publication or submission and the sequencing of box numbers. However, organizing that amount of disaggregated data into a searchable database was a daunting task. Problems of continuity, particularly in relation to the non-standardized usage of terms, were significant, as in the case of terms with political connotations (for example, "environment," "pollution," or "dust"). Finally, access to the archives was restricted legally. Under the terms of the bankruptcy proceedings, a New York law firm was empowered to determine if any of the archival material should be restricted. Many files dealing with employment histories, payment records, and hospital and health data were restricted. Many former Cassiarites are still alive; from a corporate standpoint, such restriction is a reasonable precaution. However, the legal restrictions imposed on the collection meant that much of the material dealing with environmental assessment was also restricted, while the law firm determined the risks involved in the release of specific files.

Embedded in the legal oversight process were a legal assessment of the researcher's work, the positionality of the researcher, and the potential legal hazard to the company. While restrictions exist on many archival collections, it is rare to have a third party – concerned primarily with the liability of a company it represents – to wield such control over access to archives (institutions with inherent public/knowledge value) that are directly relevant to issues of public health and potential environmental contamination. At one point, then, there were substantial obstacles to research on Cassiar. More recently, however, following extensive organizational and advocacy work by UNBC archivists, the Cassiar archive has become a model documentary and photo collection. A finding aid is now readily available online, with full descriptions of boxes and files. An extensive photographic collection of several thousand photographs, a cartographic collection, and sound and video recordings are all available. Complete environmental, corporate, and medical histories of Cassiar are all now waiting to be written.[53]

The material archival remains of Cassiar, more accessible than ever before, can be juxtaposed with a more personal and contemporary project that has emerged out of a community desire to remember Cassiar. In the years after the shutdown, after residents were forced to find jobs and create lives elsewhere, a fascinating virtual postscript to the closure of Cassiar emerged. Seeking a place to connect with lost friends and co-workers, to share memories of bygone days, and to foster the idealized sense of community partnership that characterized life in Cassiar, former residents started a website called "Cassiar … Do You Remember?" Built in 1998 by Simone Rowlinson and maintained by Herb Daum (both former students at the high school in Cassiar), the website is a rebuilt Cassiar, a site for memory and commemoration. The site is sprawling: it contains hundreds of pictures, scanned archival documents, copies of Cassiar newspapers, reminiscences of employees and residents, birth and death notices, recollections of teachers and community leaders, and a news page with links to stories about Cassiar. It offers souvenirs and a links page for other online Cassiar sources. In recent years, videos have begun to appear online: a 1960 promotional video for the mine and town, a contemporary video of drone footage from a bird's eye view, a video taken shortly after the town was razed in the early

1990s, a nostalgic photo montage of a summer of work in 1973 set to Paul McCartney's "Band on the Run" ("If I ever get out of here ..."). A notice board allows conversation between former neighbours and facilitates the planning of yearly Cassiar reunions, which take place across Canada. An "In Memory" page remembers Cassiarites who have passed on and encourages remembrances from friends and family, while a trivia section tests the user's knowledge of the ephemera of northern living.[54]

The photographs are the website's main legacy. Members have uploaded thousands of personal photos documenting the social life of Cassiar: Schmoo Daze, weddings, dances, parties, concerts, and the people who populated the gatherings that signaled community. Hundreds more photographs show the landscape, both "urban" and the surrounding area so valued by residents for its wilderness qualities. Many more show the gloomy scenes of closure, the abandoned buildings and unmaintained infrastructure sagging under the weight of neglect. Others show happier times, the men's hockey team or scenes of a curling bonspiel, a Christmas pageant or talent show, or the visit by Prime Minister Pierre Trudeau in 1968.

This informal online archive has been supplemented by a new initiative launched in the summer of 2011 seeking to link Cassiar to other modern forms of social networking. Working with the Cassiar photo collection, UNBC archives staff launched the "Cassiar Photo I.D. Project" on Facebook. The project harnesses the communal nature of social media to help identify individuals, places, and everyday activities at Cassiar that are otherwise obscured by the haphazard nature of the original accessioning. The process reads like a running historical commentary, with contributors offering suggestions and possibilities, assertions and contradictions.[55] Archivist Ramona Rose has appealed to the "intangible heritage" produced by the virtual community and its use of the UNBC networking tools.[56] The intangible heritage may be read in another way in reference to the largely elusive medical legacies of asbestos mining. The website has promoted awareness of potential health implications of working at Cassiar and advocated for former residents to be screened through WorkSafe BC. Even in its posthumous online life, Cassiar is a contested place. It is a town still being constantly reproduced, still dealing with the memory of its ruins, still toiling with a mourning for a place consigned to the auctioneer's gavel. But it is clear from its online remembrance that Cassiar engendered a vigorous sense of community that has outlasted its physical structures and transcended its material relationship with the environment.

FIGURE 1.12 Cassiar Tailings Pile

Conclusion

Cassiar was never just a dream of planners, bureaucrats, or investors; it was a full and functioning town and industrial operation, the administrative and service centre of the region and the economic heart of northwest British Columbia. Over the course of forty years, Cassiar was built into a community of almost two thousand people, with modern amenities and amusements, rich social worlds, and economic prosperity for families and single working men. The open pit mine sustained the town and the company over those forty years, but it was exhausted when the company processed all of the asbestos fibres they could extract economically. The open pit's replacement, the underground McDame mine, was undone primarily by human error and unfavourable economic circumstances. Fluctuating markets, imprecise engineering, uncertain management, and unsupportive creditors meant that McDame would not be built and mined to its full potential. The inhabitants of Cassiar and the built environment around the townsite would bear the material burden of the breakdown of the McDame mine scheme. When residents were forced out of their homes and when the material world of Cassiar was razed, an unbuilt environment of a different order was created. It contained many traces of the once-thriving settlement, which had in many respects followed the classic boom and bust cycle of northern resource towns and mines. Cassiar was built, it prospered for a time, and was then deconstructed. But this deconstruction – and destruction – was the result of the operational and business failure of the underground McDame mine. It was the unbuilt nature of a particular component of the Cassiar project that led to its demise.

The closure of Cassiar raises important questions about the outcomes of the unbuilt environment, primarily about the effects of the removal of the built environment, but also about the erasure of the human environment and its abrupt divorce from the community that sustained it. The building of Cassiar, as a town, a mine, and a corporate entity, instigated profound environmental and economic changes. The closure of Cassiar, likewise, left a vacuum of environmental responsibility. Cassiar's failure almost two decades ago has left an economic and social void. Population in the region has been halved. New industrial investments are contingent on infrastructure that is lacking; there is no substantial community to anchor it. Virtual Cassiar adds another dimension to the question of the unbuilt environment. It has little impact on the material remains of the former town, on the tailings pile, or on current mining endeavours at the site. However, as a prospective geography and as a place where past connections can maintain community futures, the site has considerable impact.

The next several chapters consider other megaproject failures. I move through an analysis of the collapse of a major transportation initiative, a hydroelectric generation plan, and a liquefied natural gas export scheme before concluding with a consideration of contemporary energy transmission proposals and their intimate connections to the mining sector and community development. Cassiar hangs above these next four chapters as a kind of test case of growth, promise, and failure. The other cases follow a similar trajectory. They overlap, both temporally and in ambition, informing the outcome and historicity of each other. Yet they remain distinct as modes of modernity and development that were never fulfilled. Together they make the unbuilt environment in northwest British Columbia.

2
Liberating Stranded Resources: The Dease Lake Extension as the Railway to Nowhere

IN AUGUST 2010, Fortune Minerals announced plans to move coal by rail from its mine in northwest British Columbia to the port at Prince Rupert. Fortune's large anthracite metallurgical coal deposit, the Arctos Anthracite Project, would remove the top of Mount Klappan and transport it across oceans. While Arctos Anthracite may seem to be the middle of nowhere, in reality it sits in the middle of a complex, peopled wilderness. The obstacles to the transportation of coal were significant, but Fortune had hit on a plan: the re-development of a "commercially competitive railway transportation option" that uses "existing infrastructure ... to rapidly capitalize on the world class resource potential of Mount Klappan and participate in the growing global shortage of high quality metallurgical coals." The existing right-of-way was impassable – even washed out in places – and upgrades would be expensive. But the new railway strategy "reduces the environmental footprint for the development, eliminates concerns for truck and port congestion in Stewart, reduces overall project risk, and provides for more rapid project execution and construction." The company would run 127-unit trains with 95-tonne car loads on the upgraded rail bed, BC Rail's Dease Lake Extension, for 150 kilometres from the mine site to Minaret, where coal would be loaded on Prince Rupert–bound trucks for shipment to "the company's potential customers in the overseas steel industry."[1]

To understand the hubris of Fortune's transportation plans, we must go back to assess a range of attempts to build roads and railways into and across northwest British Columbia over the past century. This chapter

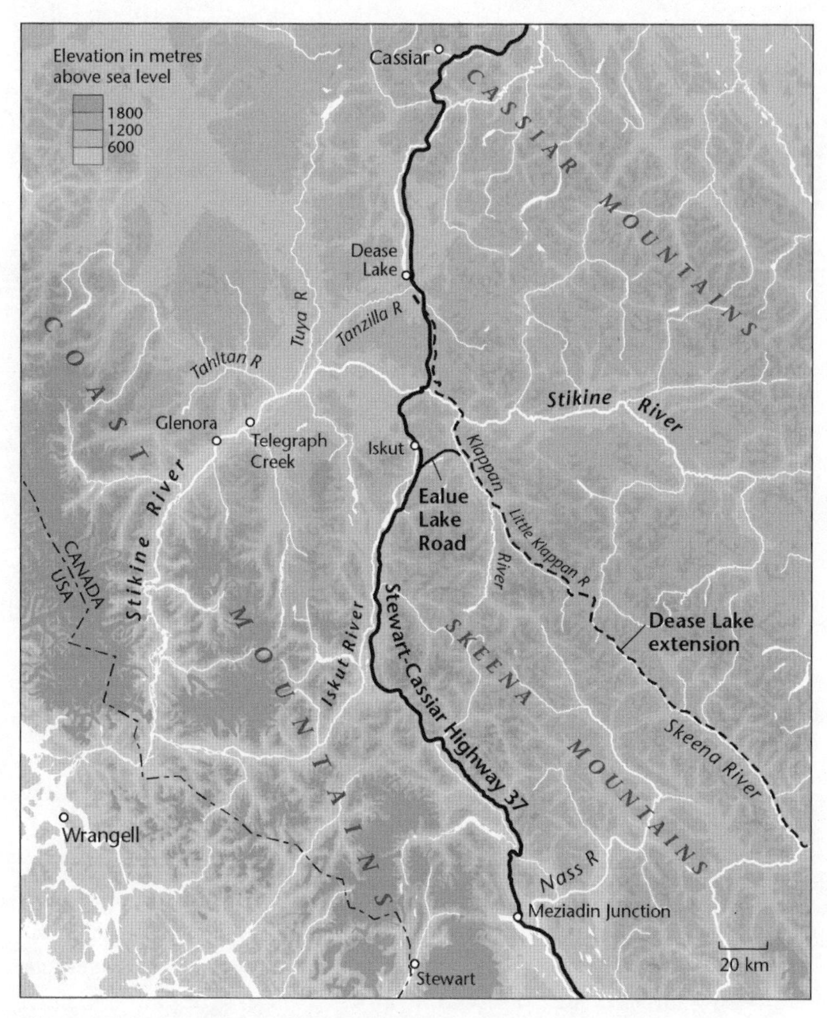

FIGURE 2.1 The Dease Lake Extension with Ealue Lake Road and Highway 37. Map by Eric Leinberger

outlines a series of efforts by companies like Fortune Minerals, governments, and private investors to construct a major transportation line to connect the north and south of the province. I focus in particular on the most spectacular failure among these schemes – BC Rail's ignominious Dease Lake Extension, a rail bed almost five hundred kilometres long, abandoned in 1977 in mid-construction amidst economic, environmental, and administrative controversy. Successive transportation megaprojects were designed to transect the Stikine wilderness, to access its hidden resources and to develop its mountains, plateaus, and watercourses. The dreamers behind these schemes prized what lay beyond the boundaries of the Stikine watershed in the Yukon and Alaska as much as the copper, gold, and asbestos that lay within. Advocates of new transportation systems envisioned rails and roads that would open up possibilities of resource extraction and movement. For politicians and promoters, Americans and Canadians, mining entrepreneurs and bureaucrats, transportation systems were the essential precursor to development. They imagined multiple different solutions to what they might generally have called the problem of access, but all of them sought to develop the north through the construction of transportation networks.

There was recurring interest in building a conduit through the Stikine throughout the twentieth century. Some ideas were rough and ephemeral; others were successful (such as a resource highway completed in 1972), realizing the goal of connecting the Stikine and its resources to the southern provincial transportation network. Still others were chimerical: of these, the Dease Lake Extension was the grandest of them all – so spectacular in its ambition and its failure that a royal commission was established to examine its entrails. Royal commission submissions variously described the Dease Lake Extension as an example of bureaucratic ineptitude, engineering clumsiness, environmental disregard, and blind political will. The state collected a surfeit of information before making its decision to shut down the project. In this case, there was no corporate culture actively seeking to protect information from the researcher. The unbuilt environment here can be derived from an open archive – though, of course, one that was partial to the powerful economic and social voices of the 1970s. State institutions lead the way for environmental reporting in the Dease Lake Extension, but they are also, through poor planning and management of a massive infrastructure project, the principal cause of the environmental degradation they sought to understand. As we will see, however, the abandoned rail grade has provided an essential line of access and has become a central axis for the manipulation of the environment in the watershed. It still functions, though not as intended and not without opposition, as an important thoroughfare through British Columbia's Northwest.

Antecedents

Early plans for transport routes through northwestern British Columbia sought to connect Alaska to the continental United States. Despite periodic rumblings and public declarations, nothing official emerged until 1928, when Donald MacDonald, an associate engineer with the Alaska Road Commission, "mapped a highway route back of the Coast Ranges from Hazelton to Fairbanks and made a tentative estimate of the mileage and costs."[2] MacDonald's surveying was done in a drafting office, with no ground-truthing, but US president Herbert Hoover and BC premier Simon Fraser Tolmie were convinced of the economic benefits of the route. MacDonald created the International Highway Association, a western transportation "advisory" group, and submitted his report to Congress in 1933.[3] With the Great Depression in full swing, it was an inauspicious time for major infrastructure development.

The report solidified American interest in the area, and surveys and engineering studies continued. The US government, anxious about the possibility of war and the perceived Japanese threat, established the Alaska Highway Commission in 1938 to examine proposed routes for what became the Alaska, or Alcan, Highway. An affiliated Canadian commission was established in the same year, although the selection of Charles Stewart (a former Liberal cabinet minister from Edmonton, whose health limited his ability to travel and work) as head of the Canadian section reflected Canadian prime minister Mackenzie King's skepticism of the scheme.[4] Both committees reported in 1941.[5] The US commission supported the coastal route; the Canadian commission offered tepid endorsement of an interior route heading northeast from Prince George.[6]

Support for the coastal route was undermined by the political maneuverings of a group of politicians, bureaucrats, and entrepreneurs eager to secure corollary business opportunities for their home constituencies on the Canadian Prairies and US Great Plains. The United States–Alaska Prairie Highway Association, with members from Alberta, Saskatchewan, North Dakota, and Montana, staged a last-minute political intervention that secured US Senate approval of the route still active today.[7] The successful route had an additional military application. It roughly aligned with much of the Northwest Staging Route, a series of airstrips and radio towers designed to provide quick response and supplies to Russia as part of the Lend-Lease Program.[8] Northwest British Columbia would be bypassed in favour of a more easterly route heading north through Fort St. John.

Instead of the long-term economic benefit of a highway, residents of the Stikine enjoyed the short-term economic benefits of another wartime infrastructure project. The Stikine River and the labour of its inhabitants were integral to the construction of the Watson Lake Aerodrome in 1941–42, just across the BC border in Yukon Territory. Because the Alaska

Highway was still under construction and could not yet be used to transport materials, machinery, and men to the north, US construction personnel decided to develop the Stikine–Dease Lake corridor as a transportation conduit to bring the necessary goods to the planned Watson Lake depot of the Northwest Staging Route. They awarded the contract in March 1941 to General Construction of Vancouver, which built and contracted boats and barges to take materials up the Stikine to Telegraph Creek. The company then improved the Dease Road between Dease Lake and Telegraph Creek (about 108 kilometres; see Figure 2.1). It hired local merchants to run trucks on the Dease Road. Others were hired to transport materials on boats on the Dease Lake–River system to Watson Lake. The improvements shortened the Telegraph–Dease trip from three days to an average of four hours. Much like previous booms connected to the gold rush or hunting and guiding economies, this new construction and transportation phase was short-lived for the residents of the Stikine. Freighting continued along the corridor through the summer of 1943. But as soon as transportation bottlenecks cleared up on the new Alaska Highway, daily life in the Stikine returned to its unhurried, pre-war state.[9]

The next major transportation initiative was instigated in the late 1950s as a major part of BC premier W.A.C. Bennett's bourgeoning northern development dreams.[10] Bennett's business courtship of Swedish industrialist Axel Wenner-Gren resulted in the imagining of the Pacific Northern Railway (PNR), a track roughly similar to the coastal route envisioned by Donald MacDonald.[11] Construction was to start in the summer of 1960 at an estimated cost of $400 million.[12] But delays followed. British Columbia Minister of Transportation Phil Gaglardi reported in the summer of 1960 that the Wenner-Gren "company proposes a 700 mile rail line ... on a route that has been extensively surveyed at a reported cost of several million dollars [while] estimates as to total costs including construction, rolling stock and financing are $250,000,000."[13] The cost discrepancy was not explained. Much like Wenner-Gren himself, the PNR directors were cagey about their plans, short on specific information, and cautionary with promotional material. There is no doubt, however, that Wenner-Gren gained considerable economic concessions because of his promotional activities. In addition, the planners behind the PNR anticipated linking up with the Alaska rail network to create a true transcontinental north–south rail network.[14] But the Wenner-Gren railway dream, and its financing, surveying, and touted infrastructural advantage, proved as tenuous as the other schemes to move goods through the Stikine. By the beginning of 1962, the PNR was abandoned.

The only successful transportation scheme to transect the Stikine was initiated at roughly the same time as the Wenner-Gren initiative. Construction of the Stewart-Cassiar Highway began in 1958. Jointly funded by provincial and federal governments under then Prime Minister John Diefenbaker's "Roads to Resources" partnership program, which was designed to spur northern infrastructure development, the highway was intended to link Stewart to the Alaska Highway just west of Watson Lake within five years.[15] Yet aside from some small-scale logging, the only potential commercial user of the road was Cassiar Asbestos, located 550 kilometres north of Stewart but anxious to reduce its transportation costs from its remote location. In fact, the Cassiar Asbestos Company had already upgraded the ninety kilometres of road between the mine and the Alaska Highway at Watson Lake in the early 1950s. The new highway, built on top of the Cassiar Road, encountered many of the same problems as previous road-building efforts. Depending on the time of the year, mud engulfed trucks, new road sank in soft soils or washed away in flash floods, and frozen ground could not be transformed into gravel. Supply lines were long and expensive. The last section was finally completed, wildly over budget, in November 1972, and a connecting all-weather route between the Stewart-Cassiar Highway and Hazelton was opened to the public the following year.[16]

Communities and individuals in the Stikine had little say in route selection for any of these projects. From the late-1940s to the early 1960s, Indian Agent Harper Reed promoted a Pacific highway, emphasizing an anticipated intersection with Atlin, his home base, approximately 240 kilometres northwest of Dease Lake.[17] After construction on the Stewart-Cassiar Highway began, Reed focused his energies on the promotion of spur roads, designed to connect communities on the highway with the towns, resources, and ports of southeast Alaska. Reed's local interest was shared by politicians from Alaska, the Yukon, British Columbia, and the Pacific US states. These efforts came to a head in the summer of 1960 in Victoria at the Alaska-Yukon-British Columbia Conference attended by high-level politicians from these governments. Each presented transportation prospectuses, largely focused on integration and mutual economic benefit. Spur roads to Juneau, Ketchikan, Wrangell, and other coastal towns were promoted but proposals failed to yield definite construction commitments. The abandonment of the Wenner-Gren rail project signaled the end of another possible tie-in partnership. With few business enterprises beyond Cassiar Asbestos to use and promote infrastructure development, and with the continued incremental construction of the Stewart-Cassiar Highway, the conversation about spur lines and coastal routes died away. These northern extension discussions were revived several years later under a different guise: the BC Rail Dease Lake Extension.

The Dease Lake Extension

> *There is no good place, but this is as good a place as any to relate the sad story of over-runs and under-estimates, very expensive field-redesigns, massive expenses incurred to satisfy environmental authorities, miscalculation and litigation.*[18]
>
> *Royal Commission on the British Columbia Railway*

The Dease Lake Extension was actually three construction project phases linked by a simple notion that tied provincial economic growth to northern infrastructure expansion: the Fort St. James Extension, running 117 kilometres from Odell to Fort St. James, completed in 1973 for just over $18 million; the Takla Lake Extension, running 130 kilometres farther north between Fort St. James and Leo Creek, completed in 1973 at a cost of $23 million; and a third section, a 540-kilometre line commonly referred to as the Dease Lake Extension, begun in 1970 and slated for completion in 1974. When construction of the extension was suspended "temporarily" in March/April of 1977, all 540 kilometres of rail bed had been built, but only 253 kilometres of track had been laid and only 84 kilometres of this was operating, between Leo Creek and Bulkley House.[19] Construction costs had increased exponentially. Approval of the line in 1969 carried an estimated construction cost of $68.9 million, but the actual cost had risen to $360 million at the time of abandonment.[20] In February 1977, while the provincial government under W.R. Bennett was attempting to suspend work on the extension without penalties, a provincial royal commission was struck to assess the economic potential of a rail line into northwest British Columbia and the feasibility of continued construction of the extension. This was to be a cost-benefit exercise, designed to quell public unrest at the increasingly obvious economic and environmental debacle. Commissioners were to decide if the dire economic realities of the Dease Lake Extension should result in its abandonment or whether the economic possibilities in forestry and mining envisioned along with its completion were enough to warrant its continuation.[21]

Submissions to the royal commission exposed the managerial inadequacies, environmental disregard, financial manipulations, technical clumsiness, and political prerogatives that motivated construction and led to the failure of the extension.[22] Contributors catalogued the litany of abuses and oversights, shortcuts and cover-ups perpetrated by contractors, managers, and under-qualified railway personnel. The picture that emerges reveals a corporate and political culture of northern development, geared towards the production of energy and the movement of goods but unprepared to deal with the environmental and economic conditions of social and geographical marginality so familiar to residents of northwest British Columbia. The Dease Lake Extension was abandoned, not because it was a bad idea doomed to fail, but because planning and construction practices were deficient and there were insufficient mechanisms for oversight and change.

The development ideals that underpinned the Dease Lake Extension were based in the "northern vision" of W.A.C. Bennett, intended to "herald in an era of renewed and torrid development ... a modern day gold rush."[23] Emboldened by the success of other northern development megaprojects, such as the W.A.C. Bennett Dam near Hudson's Hope, Bennett returned in the late 1960s to his earlier northern transportation goals, with a bold plan for railroad development. This was to be a "provincial development tool," brandished in negotiations with the federal government and used as a carrot to investors and geologist-prospectors leery of transport costs associated with any major resource finds. Bennett had more than a hint of the evangelist when describing his plans, and in lamenting lost opportunities: he had "a vision of opening the vast untapped British Columbia north; a vision of providing our American neighbours a vital link between the States of Alaska and Washington; and, a vision of a new Yukon."[24] The Dease Lake Extension, a "pioneer railroad," would ensure the economic vitality of northwest British Columbia and have considerable effects in the southern parts of the province eager for power and resources.

The Dease Lake Extension was part of a larger provincial strategy to encircle with rail an area of uncertain but potentially lucrative resources in the northern half of the province. Along with the Fort Nelson Extension in the east, the Dease Lake Extension would provide links to the rest of the continent and, through improved port facilities, to overseas markets. Furthering this dream, Bennett envisioned the creation of a resource traffic corridor to Alaska and the Yukon. Railways would stimulate resource exploration and commercial enterprise. Aside from the Groundhog coal fields and the Cassiar Asbestos operation, northwest British Columbia had no commercially viable resource opportunities and few confirmed deposits. In Bennett's reckoning, however, the completed rail link would encourage exploration and underpin the economic prospects of identified copper, gold, and forestry resources. Construction would follow a "minimum standards, maximum speed" principle. The Dease Lake Extension would be built as quickly and cheaply as possible and would only be upgraded when traffic justified further investment.[25]

Bennett had some institutional backing from the federal government for the Dease Lake Extension. The federal Ministry of Transport contracted the consulting firm of Hedlin Menzies in the late 1960s to analyze transportation options and "the complete spectrum of the region's overall mineral and forestry resources" for northwest Canada (Yukon Territory and the northern half of British Columbia).[26] The firm saw dramatic resource potential for the region as a whole, though this potential was contingent on an expansion of transport facilities. But Bennett wanted a firm commitment from the federal government. Instead, he perceived vacillation and resolved to begin without federal assistance. BC Rail had compiled more specific locational studies in the wake of the failure of the Pacific Northern Railway. A BC Rail–commissioned study released in December of 1969 as a buttress for the newly conceived Dease Lake Extension noted regional potential in agriculture, recreation, forestry, oil, and gas as well as hydroelectricity. Although the section on mining was positive, this study, too, noted that "the majority of known finds are uneconomic due to inaccessibility."[27] The message was clear: in the parlance of industry, these were "stranded resources." The resources were probably there, but better infrastructure was needed to find out how much was there, how best to extract them, and whether such extraction was economically feasible.[28]

A change in government, from the populist Social Credit Party to a progressive NDP in 1972, did not change northern railway development policy, although the NDP alleged serious irregularities in the financial management of earlier northern railway construction. An analysis by accountants Touche Ross and Company revealed that previous BC Rail directors (including Bennett and Minister of Lands and Forests Ray Williston) had misrepresented the financial details of the Dease Lake Extension. Still, the Dease Lake Extension was considered financially feasible if a federal capital cost grant (already agreed to in principle) could be secured and several other mitigating factors were met.[29] At around the same time, reports of the negative environmental consequences of the extension were beginning to circulate. These reports were the first substantiated evidence of the economic and environmental mismanagement that eventually halted construction.

At base, the problems of the Dease Lake Extension were attributable to the route that had been selected. Time and again, the Royal Commission heard that the railway had been placed in the "wrong location." The original survey tactics employed by Ray Williston, and the general lack of advance planning in the late 1960s, were at the root of the location problem. Williston had selected the route himself. He flew in a helicopter along the general line of the abandoned Pacific Northern Railway of the Wenner-Gren group. He did have some surveying experience but he did not use it on the ground, apparently preferring to study general lines on large maps in his Victoria office. No engineering and geological studies were conducted. Route location was not chosen in order to be close to known resource areas. The route simply went north. It was left to local contractors to move it in that direction as quickly and cheaply as possible. According to the royal commission submission made by the Fish and Wildlife Branch of the BC Ministry of Recreation and Conservation, unnecessary ecological damage occurred during the building of the railway because of "the lack of any workable process at the conceptual and design stage of the railway to incorporate on a planning basis reasonable environmental concerns."[30]

Problems encountered in building through the Bear-Sustut River system are illustrative of larger issues. In a rare display of sound environmental judgment, the fisheries branch of the federal department of environment required BC Rail contractors to relocate the rail bed away from sensitive fish spawning habitats along the Bear and Sustut River flats. Contractors had been working directly beside the rivers – sometimes even on top of them. They were now forced to move the line away from the riparian zone, into very difficult, and consequently very expensive, hillside terrain.[31] The Kitimat-Stikine Regional District advocated shifting the line towards the Nass, Bell-Irving, and Iskut valleys, where, they said, forestry and mining potential was greater and access to other northern centres, like Terrace and Prince Rupert, was easier than along the chosen route. The regional district had been pushing this case since 1973, but BC Rail and the provincial government were reluctant to change the route in spite of economic benefits and lower capital costs, believing that any federal grant for the extension was tied to the original plan.

The most concrete environmental effect and the greatest economic cost of the Dease Lake Extension were due to the movement and disposal of gravel and other construction material. Gravel, a seemingly benign foundational construction technology, has far-reaching consequences in riparian habitats, on larger interconnected ecosystems and on local economies that rely on rivers for many uses. In its discussions of the Bear–Sustut re-routing activities, the royal commission stated that additional excavation necessitated by the fisheries decision had resulted in "increases from 1,200,000 cubic yards of common material to 5,477,000 cubic yards and from 30,000 cubic yards of solid rock to 506,000 yards of solid rock plus 128,000 yards of loose rock. At least 80% of the material excavated had to be wasted rather than being used as fill in the balancing process."[32] The problem of gravel and rock movement, or infill, affected every aspect of construction on the extension. The downstream effects of this problem were profound.

The lack of engineering enterprise, both before and during construction, was a direct result of the inadequate surveying practices employed by Williston and BC Rail. Submissions often mentioned the lack of "pre-engineering," the surveying and assessment work conventionally required in large infrastructure projects. The BC Rail engineering department was too small, poorly trained, inexperienced in rail engineering, and pressured by the directive to work quickly. Nelson Hepburn, a consulting engineer to the royal commission, attributed the lack of pre-engineering to BC Rail's goal of minimizing costs and to a dearth of qualified personnel. Hepburn claimed:

> The railway was thus in an extremely difficult situation when the decision to proceed with construction was given, and was faced with the choice of either enticing experienced engineers from the two continental railways – or training their own as best they could during the progress of work, all for a construction program which at the time was visualized as lasting for only a few years, with no evidence of ongoing work ... All of this could have contributed to the fateful decision to proceed to contract without the implied proper pre-engineering. The BC Rail ... failed to draw on skilled location and construction engineers with the experience that could have been adopted easily to meet their needs.[33]

BC Rail engineers largely agreed with this assessment but were powerless to do something about it because of the organization's intransigent managerial structure. Stanley Oakes, an engineer with twenty years of professional experience, began working for BC Rail in 1969. While still employed by the company, he wrote to the commission that he felt compelled to "bring to light and to emphasize the fact of the lack of adequate involvement of Professional Engineers during the early stages of the planning, design and construction of the British Columbia Railway's Northern extensions. Many of the problems which have subsequently developed are attributable to this fact." Oakes suggested a correlation between engineering practices and environmental problems, implying culpability on the part of BC Rail:

> Less publicized has been the fact of the considerable number of complete or partial failures of embankments, drainage structures and the like which have occurred. These failures reflect badly on professional engineers within the Company who might reasonably be expected to have been closely involved in the work ... the amount of engineering involvement at the professional level was extremely limited and fell far short ... of that which should have been properly provided. I expressed my concern about this to the management of the Company on a number of occasions with little result and it was only after a number of construction failures began to develop that there came to be a gradual recognition and acceptance of the need for more engineering.[34]

Royal commissioners pointed to in-field revisions necessitated by the lack of pre-engineering. Again, this had a direct bearing on the amount of material that had to be moved. Contractors had to deal with the decision to "increase base widths of cuts from an 18-to-20 foot standard to 32 feet in solid rock and 24-to-26 feet to 54 feet in other materials."[35] In some cases this had to be done to maintain track integrity and stability, and in others cases it was simply necessary to allow on-track snow plows to function properly.[36] Sometimes these decisions were made with managerial approval and sometimes they were made on the spot. Consideration of cost increases appears to have been negligible.

A report by engineering consultants Swan Wooster in 1973 on engineering and contracting practices generally backed up the assertions of the professional engineers. The report claimed that engineering on the extension was largely satisfactory, with the important exception of the pre-engineering, which was substandard at best. The authors took issue with BC Rail's "current practice [which] is to perform almost all detailed engineering of earthwork in the field, concurrently with construction. This contrasts with the more normal practice of executing most of the detailed engineering on any given segment of the line prior to the award of a construction contract."[37] This irregularity led to low earthwork estimates. The hiring of a geotechnical engineer and a senior design and construction engineer would have proved invaluable in public relations and, by their being able to provide advance notice of railway activities, in the expression of transparency with other public agencies. Because there were no reliable earthwork estimates, there was no yardstick to check performance. According to Swan Wooster, better pre-engineering was simply good business: it would have paid for itself.[38] The BC comptroller-general agreed. Asked by Barrett's NDP to review BC Rail finances, the comptroller-general stated in April 1973, before the full extent of the Dease Lake Extension failure was known, "with respect to the awarding of contracts for rail line extensions I confirm that the company's officials have seen fit to limit spending on preliminary engineering studies, a practice which directly contributed to overruns in the order of $17,000,000 on contracts completed to date."[39]

Responsibility for the shoddy engineering can be attributed to BC Rail managerial decisions and the pervasive problem of infill, but blame for the overruns can be shared by the consulting and contracting practices employed in the construction of the railway. Every level of construction was sub-contracted, and BC Rail awarded each contract to the lowest bidder, regardless of experience and qualifications. This was in keeping with Bennett's "minimum cost/maximum speed" principle but had predictable ramifications. From the outset, the Dease Lake Extension was beset with cost overruns and contracting difficulties. Zanatta and Levae Bulldozing, which won the first contract in August 1970 for clearing the right-of-way, made claims for additional remuneration because of problems getting men and equipment to the job site.[40] On the same section of the extension, Catre Industries exceeded its contract by $4,489,017.27. The rock quantities that it had to move were double the original estimates, which was "attributable to the fact that the line [chosen by BC Rail] closely followed the shoreline and during construction it was found that the rock shelves in which the grade had been projected were completely unstable to carry the railway grade."[41] Fourteen such revisions had to be made on this contract alone. Catre had to sue but was eventually paid 137 percent of the original contract price. Similarly, Keen Industries claimed an overrun of $4 million because "the quantities were not accurately estimated from the engineering and field information available at the time the contract was let."[42] Keen was paid a 100 percent overrun. The royal commission commented on the case of KRM Industries, which had its contract terminated after $18,322,047 had been paid on a job that had been awarded at $7,448,399. As happened in most cases of contract dispute in the Dease

Lake Extension project, the KRM contract was settled only after a lawsuit.[43] These kinds of overruns and legal costs were chronic.

However, the problem of gravel quantity and earthwork was only the most materially obvious reason for the massive cost overruns. The general rate of inflation in the 1970s was the biggest blow to contractors.[44] There was no mechanism to adjust the value of contracts for inflation, so contractors took the brunt of the inflationary hit when their contracts took longer than expected because of the extra work involved in moving extra material. Contractors could not properly deal with escalating costs within the structure of their static contracts. As inflation worsened, "the cost of getting northern BC resources out went up, while the price obtainable for them in world markets went down."[45] This had far-reaching consequences for the economic viability of the extension as a whole, as it constricted its customer base and put additional pressure on the potential customers of the companies that were moving goods on its line. Contractors also took huge economic hits when the line was abandoned. Chinook Construction and Engineering had been awarded the contract for grading and culvert installation between miles 286 and 308 of the Dease Lake Extension. The contract was worth $10.3 million, of which an estimated $5.28 million had been completed when the company was ordered to pull all personnel and equipment off the job. In its royal commission submission, Chinook suggested that construction should continue because of the danger of damage to existing work, the larger cost of later completion, and the potential loss of economic benefits to the province through the loss of eighty jobs and the government's loss of $2.085 million in taxes. Presumably, the $5 million of lost work and the corresponding company revenue were part of the company's deliberations as well. Contractors were able to charge more than expected for their work, but they often faced considerable difficulties in completing the work and in securing payment.

Although Swan Wooster reported that construction practices on the Dease Lake Extension conformed to "industry standards," many other commentators identified serious and avoidable environmental problems associated with the building of the line.[46] The BC Fish and Wildlife Branch lamented the "flagrant disregard by the major contractors for even the most basic environmental concerns and the inability ... of BCR to exercise proper control over their construction practices." This was aggravated by "the complete lack of any educational program by the Railway or by government to acquaint workers with these concerns."[47] The branch submitted its own plans for encroachments and diversion to BC Rail staff, but was continually frustrated by the lack of input and by the low quality of environmental stewardship during construction near streams and rivers. In 1972, L.P. Levernan, vice-president of the BC Wildlife Federation, travelled along the Dease Lake Extension on a reconnaissance trip. He complained of companies using bulldozers instead of shovel cranes to move gravel, a situation he compared to the damage done by placer mining in juxtaposition to the minimal impact of panning. Far away from one construction camp and the railbed itself, Levernan found that construction personnel had destroyed the muskeg by "playing games" with their vehicles.[48] Four years later, David Bustard, the habitat protection biologist with the BC Fish and Wildlife Branch, claimed that within the previous year, extensive unstable earth cuts had been opened up in the Upper Skeena, Spatsizi, Tanzilla, and Stikine Rivers. Older cuts on the Klappan, Little Klappan, Sustut, and Bear Rivers had still not been adequately secured. In addition, approvals granted for waste disposal areas and stream crossings based on the premise that proper stabilizing would be carried out had been left partially completed and unstable. Programs designed to stabilize numerous cuts and fill situations with proper drainage structures and revegetation had also been abandoned, only partially completed.[49]

Engineering problems as well as poor contracting principles and destructive construction techniques created an untenable situation, where environmental damage became an almost inevitable outcome. Poor management compounded the problem.

The BC Construction Association defended its members and pointed to a culture of deference towards engineers in the corporate culture of BC Rail. It also suggested that the royal commission's terms protected the engineers and planners of the extension from culpability. The BC Construction Association thought it was "amazing, considering the omnipotence attributed to the engineer in these clauses, the lengths that the Conditions go to, to ensure that the Company is protected from the consequences of errors made in design and specifications by the Railway's engineering staff." Even more galling was that the "General Conditions are filled with penalties which the Company can impose on the contractor, but there is no remedy provided for the contractor in the event that the Company does not meet its obligation."[50]

Robert Dhensaw, a construction foreman with BC Rail, complained about the lack of internal supervisory promotions from within the ranks of the BC Rail workforce, particularly as more managerial promotions seemed to come from outside company ranks. The lack of training programs was a particular cause for concern. Dhensaw criticized the use of company vehicles for non-company purposes, both by executives and by personnel engaged in the destructive activities alleged by Levernan. But Dhensaw's greatest scorn was reserved for the ineffective managerial culture and very poor management–labour relations. According to Mark Crawford, an analyst with the BC Project in the political science department at the University of Victoria, the challenges and problems of the Dease Lake Extension were "directly related to the authoritarian style of decision-making in which employees, railwaymen, engineers and economists were required to carry out management policies and never help to formulate or even question them."[51] The authoritarian style went further. Poor planning and minimal public relations were "compounded by poor relations between the BCR and all of the most important private sector groups directly involved in, or affected by, the construction of the railway: Keen Industries Ltd., MEL Paving Ltd., the union, the native community and other groups whose land was expropriated or devolved."[52] Constructive leadership was clearly lacking at BC Rail. This leadership failure had material consequences along the Dease Lake Extension corridor.

As the lack of surveying and engineering and the construction practices of contractors became more widely known, assessments of specific environmental problems associated with railway construction began to surface. The BC Fish and Wildlife Branch asserted that the "construction of a project as extensive as the Dease Lake extension poses many potentially severe environmental impacts some of which may be readily apparent while others are more subtle but insidious in their effect."[53] The problems were legion and mostly related to the movement and disposal of gravel, especially in riparian zones. Broadly speaking, the two major problems were erosion and habitat loss. Both were the result of maladapted construction technologies, and their effects were exacerbated by the abandonment of the extension.

Siltation and sedimentation resulted from work on the banks of watercourses. The roe and spawning habitat of salmon, trout and other fish were affected as gaps between rocks and gravel were filled in, and water percolation and oxygen content decreased. High water turbidity can block fry emergence from spawning beds by reducing light penetration, which stunts plant life and the production of food for fish. Siltation occurs naturally but also may be caused "by heavy machinery working in a watercourse, failure of culverts, drainage from unsettled roads and ... erosion from unstable banks, cuts and channels."[54] According to David Bustard, mass wasting was exacerbated by poor construction practices. Bustard observed that "slumping from cut and fill slopes is occurring extensively along the grade from the Upper Skeena right through to Dease Lake. Material from these slumps may go directly into fish streams or indirectly via small creeks and ditches, causing siltation of spawning and rearing areas."[55] His report on the environmental problems of the extension, a copy of which was submitted to the royal commission, contained photographs and commentary on thirty-two examples of various adversely affected areas. Explaining a photograph of the Upper Skeena River, Bustard wrote:

> Waste material has been pushed to the side of the rail grade and has oozed down into the Upper Skeena River. An extensive section of the grade in the vicinity of the Gunanoot Camp is similar to this ... Why hasn't this material been end-hauled to a safe site for disposal? It is our understanding that the contractor, if instructed, could be requested to back haul material without additional cost to BC Rail. Time and Again, this material has been pushed to the side and left to ooze into the rivers.[56]

Another photograph shows significant surface erosion and slumping on the Stikine River: "Prior to abandonment, BC Rail was in the process of benching the slopes, installing cross-drainages and revegetating these cut slopes."[57] Abandonment of the extension meant that any remediation projects were left incomplete.

Erosion and siltation problems were particularly serious around culverts, which were cheaper than bridges but carried a heavier environmental burden because they reduced channel size, affecting downstream water velocity and upstream water storage. Levernan was emphatic in his opposition to culverts, claiming that regardless of construction specifications culverts would always adversely affect the ecology of a river or stream in ways preventable by a larger bridge structure.[58] Culverts can be barriers to fish migration due to their installation height and interior velocity, and they can produce scouring of riverbeds below the culvert. Warning against abandonment of the extension in its submission to the royal commission, Chinook Construction noted the possible environmental repercussions of an unfinished culvert, "covered with a minimum coat of material" that "could easily be washed away" in a bad winter "leading to the loss of the culvert as well."[59]

The extensive manipulation of streams and rivers also resulted in channelization which could take the form of "the straightening of a stream, the levelling or widening of a streambed, relocation, realignment, ditching and gravel removal."[60] This is in part what occurred on the flats of the Bear and Sustut Rivers. Channelization has many effects, not least of which may be increased flow and attendant scouring and bank erosion. Pools, side channels, and riffles are often essential rearing, feeding, and resting habitats for fish and may be destroyed or altered by channelization. The removal of streamside vegetation results in rapid bank erosion and can alter the temperature regime of the river by the introduction of silt and other materials and by the reduction of the shading effect. In addition, energy inputs can be reduced through the loss of leaf fall, insects, and other dissolved nutrients. Aside from the poor general productive health of the river, these cumulative effects caused serious ramifications for spawning salmon in the upper Skeena and for interior fish living in other connected rivers.[61]

Any consideration of the potential environmental effects of the Dease Lake Extension is necessarily speculative because of the lack of inventory data created before construction – as environmental organizations consistently pointed out. There was almost no pre-construction inventory of fisheries resources. Most of the detailed knowledge about wildlife was held by long-term residents, trappers, Indigenous peoples, guides, and prospectors, but these people were not consulted, even though many of their livelihoods were greatly affected. Indeed, the environmental assessment apparatus was hamstrung by a number of institutional factors that limited its effectiveness and reach, diluted its message, and hampered its ability to change contractor practices. Organizations like the Fish and Wildlife Branch and the BC Wildlife Federation were held back by inadequate funding for enforcement and monitoring agencies.[62] This led directly to the lack of oversight, inspection capability, and overall management of wilderness areas. Agencies simply could not travel to and provide comprehensive assessments of all the areas for which they held regulatory responsibility. Faced with mounting concern from environmental assessment organizations, BC Rail appointed a guardian on the extension to carry out the required monitoring. The guardian's effectiveness was mitigated by the size of the area to be patrolled and the fact that he was restricted to one flight per week, the same flight taken by BC Rail construction staff.[63] This reveals a general lack of capacity and the disordered mandates of the monitoring agencies responsible. Overlapping jurisdictions present the possibility for coordination but, arguably, more often result in disorder. In the end, the BC Fish and Wildlife Branch cited, as a major contributing factor to environmental disturbance,

> the lack of capability ... to monitor construction activity in any but the most cursory fashion. The logistics of patrolling the line without adequate

> manpower available defeated any attempts at consistent surveillance. We must also admit that the monitoring services which were provided by the Branch did not encompass sufficient construction and engineering expertise to provide the type of professional guidance necessary.[64]

Problems of expertise, capacity, and environmental knowledge reverberate along the Dease Lake Extension to this day as locals and environmental assessment personnel struggle to control who uses the extension and for what purpose.

Increased access to the area for hunters and recreationists may be the most profound environmental legacy of the extension. Abandoned and unmonitored, the railbed allowed vehicular access to an area previously only accessible by helicopter or float plane, on foot or on horseback. There was an immediate increase in hunting activity, with the corresponding pressure on wildlife populations. Again, there was no real inventory of wildlife data apart from the knowledge held by locals. But general experiential knowledge painted a grim picture for diverse animal populations, which became new recreation opportunities for hunters. The BC Fish and Wildlife Branch asserted that "within a 20 minute walk from the rail grade along the Klappan River, one can be amongst stone sheep and mountain goats." Moose were common along the entire stretch of the extension, from Takla to Dease Lake. Eighty kilometres of the extension ran alongside the Spatsizi Provincial Park (now the Spatsizi Plateau Wilderness Provincial Park), established in large part because of its outstanding wilderness values (Osborne caribou, moose, grizzly and black bear, stone sheep, mountain goat, and timber wolves).[65] Wildlife management personnel expected an immediate influx of recreationists eager to exploit the newly accessible area and feared their inability to manage the influx. In 1976, a 200-kilometre no-hunting area was imposed by the BC Fish and Wildlife Branch along the Dease Lake Extension in an effort to protect vulnerable wildlife populations. Some inventory work was accomplished but hunting returned to normal restrictions in the following years.

Wildlife also suffered increased pressure from expanding mining and forestry activities. The start of construction brought a surge of mineral exploration. This was still speculative, but it was hoped that deposits in the region would be richer than those in areas easier to access.[66] There was some discussion, both for and against, on using the extension as a resource road. The BC Wildlife Federation wrote about the dangers of a resource road, claiming it was self-evident that if there were no resources for a railroad to carry then there would also be no resources for road traffic.[67]

But resource companies were adamant about the inherent and forward-looking value of a resource access line. Stikine Copper held a massive copper-gold tenure claim approximately 120 kilometres west of the Dease Lake Extension, known as Galore Creek. For Stikine Copper, an economically viable access route was the main stumbling block to the profitability of its enterprise. In its estimation, though the company would not commit to using it, the extension could provide an essential access route for the transportation of ore to port. Stikine Copper stressed the potential economic benefits to the province of a successful mining operation: nine hundred jobs, a new townsite and tax base, corollary transportation options with other area resource companies, and tax and other corollary revenue from the estimated mine development costs of $400 million. The company even advocated the possibility of a series of spur lines, including one heading towards the mine site, though as it conceded:

> It is recognized that this suggestion for another costly extension comes at a time when some are questioning the wisdom of extending even as far as Dease Lake; however, it is made in the spirit of accelerating the development of northern British Columbia and the adjoining Territories, thus improving the economics of the entire northwestern line. It is further recognized that if this extra extension were to be implemented, it would also serve the interests of Stikine Copper Limited in a way which we believe is reasonable.[68]

Cassiar Asbestos agreed with Stikine Copper. The extension was an "exciting prospect" and its "basic opinion" was that it should be completed. The company also would not commit to using the extension, but it looked forward to analyzing the economic opportunities that might present themselves with the completion of the line:

> Certainly our asbestos tonnage from Cassiar would be shipped via a Dease Lake extension if it were built providing the rates were competitive with other transportation facilities ... [but] it should be considered as a road to resources, and in order to foster economic development it should offer attractive freight rates to industry and business that either exists now or will exist in the future.[69]

Increased access was simply a competitive advantage for resource companies, to be exploited much like one would mine ore from the ground.

In recent years, the extension has been used by resource companies (including Shell, Fortune Minerals, and Klappan Coal) to move personnel

FIGURE 2.2 Klappan River Bridge (Summer 2010), Where the Dease Lake Extension Meets the Ealue Lake Road.

and equipment to mining camps; it has provided access to the area for the past forty years. In particular, the length of the extension stretching 125 kilometres south of the Stikine River is under pressure from resource companies. This section runs parallel to the Stewart-Cassiar Highway but does not intersect the larger road. These companies are able to access the rail grade by vehicle because of an access road built to facilitate the delivery of BC Rail materials and men to the job site between mile 190 and mile 286. This twenty-six-kilometre access road, now known as the Ealue Lake Road, leaves the highway south of the village of Iskut and meets the extension forty kilometres south of the Stikine River, just after crossing the Klappan River. The Ealue Lake Road and Klappan Bridge were built by Keen Industries, which at the time was also working on the Stewart-Cassiar Highway.[70] The Ealue Lake Road provides year-round access for resource companies and hunters to the extension, which restricts access for unapproved vehicles, but is essentially unpoliced. (There is a sign at the entrance of the road, warning users of such restrictions: it has five bullet holes in it.)

The highway turnoff to the Ealue Lake Road is at a nondescript gravel quarry. Behind the quarry is a Tahltan hunting/outfitting camp that has

FIGURE 2.3 The Dease Lake Extension, Summer 2010

been used intermittently as the base for blockades set up to protest mining and hunting activity along Ealue Lake Road and the Dease Lake Extension, and in the Sacred Headwaters, which lie adjacent to the extension, starting at a point about eighty-five kilometres south of the Stikine River. The relations between access, exploitation, and conflict have been facilitated by the construction of the Dease Lake Extension. Shell was repeatedly obstructed in its attempts to drill coal bed methane wells, and has recently decided to relinquish the extractive potential of its claim, in large part because of activism by locals, especially members of the Klabona Keepers Elders Society, and a broader cohort of metropolitan environmentalists, including Wade Davis and David Suzuki, who both have long-standing personal attachments to the Stikine. Fortune Minerals has a better record but it, too, has faced Tahltan concern most recently in the summer of 2013, when roadblocks were set up, and the company was forced to postpone summer extraction plans (and eventually made the decision to sell its interest). Red Chris, a copper and gold mine located just off Ealue Lake Road at kilometre 6, has a poor record in dealing with local communities. In January 2010, it was sanctioned in a ruling by the Supreme Court of Canada for lax environmental assessment standards.[71] The Dease Lake

Extension may have been abandoned, but the cascading effects of access continue to shape human–environment relations in the Stikine.

The problem of access relates directly to the socioeconomic conditions of people in the Stikine and in northwest BC in general. Several studies trumpeted the large-scale economic benefits of the extension. The Touche Ross report, so integral to the NDP's decision to continue construction in the face of mounting evidence of the extension's economically perilous situation, estimated that 5,500 jobs would be created in northwest British Columbia, and between 10,000 and 17,000 jobs would be created in the province after the completion of the extension.[72] Jean-Paul Drolet, federal assistant deputy minister in the department of energy, mines and resources, claimed in 1976 that once transportation conduits were established in northwest British Columbia, new mines could employ 4,500 people by 1990. Mineral processing plants would employ a further 2,500 people, and 14,000 would work in secondary industrial sectors related to mining, with equally high tertiary employment figures to supply the industrial needs.[73] Locally, the extension developed increased opportunities for entrepreneurs who used the environment in their businesses. Increased access meant more hunting, guiding, and outfitting business; it took less time on the trail to get to and from preferred destinations. The increased business came at an environmental cost, however, and some outfitters had their territory bisected by the rail grade. But many guide outfitters expanded their operations and also developed low-impact business models around trail rides, photography, mountain climbing, and canoeing and rafting. Local businessmen were also given incentives to bid on construction contracts. Edward Asp, a Tahltan businessman, established Dease Valley Resources and employed mainly Tahltan workers. The company ran into cost overrun and earthwork issues similar to those experienced by most other contractors.

There were no comprehensive studies of the socioeconomic effects in the Stikine. There was, however, some socioeconomic analysis further south within Carrier-Sekani communities to the north of Fort St. James. This work was instigated by a blockade set up by the Stuart-Trembleur Band (now the Na'Kad'zli First Nation) in the spring of 1975. The blockade across the extension went up on April 28 in order to force the band's demands for compensation in exchange for the 378 acres that had been expropriated by BC Rail to build the extension. The Stuart-Trembleur band was asking for $7 million dollars and a 3-to-1 land exchange. BC Rail agreed to the land exchange but offered only $50,000 as financial compensation.

In late summer, the Stuart-Trembleur band lifted the blockade and negotiations continued.[74] Slightly further north, the Takla Lake band (now the Takla Lake First Nation) had to deal with the environmental disturbances to the Bear–Sustut Rivers system. In 1972, there were major mudslides in the area, with serious effects on salmon ecology. The Takla Lake band was assured by BC Rail that the area would be cleaned up and no further disruptions would take place. The following year, the band reported to the BC Wildlife Federation that things had gotten worse, including excessive garbage and chemical pollution. The federation sent a delegation to the area and confirmed reports of bank erosion, slumping, and streamside dumping. The federation used its institutional heft to force BC Rail into taking action. It was pleased to report the following year that improvements had been made. But the Takla Lake Band still had to deal with the long-term environmental consequences of river pollution and flow disruptions, which were especially onerous dislocations considering the socioeconomic importance of salmon to the community.[75]

Conclusion

On August 25, 1978, the royal commission described the government's decision to suspend construction of the Dease Lake Extension as "inevitable and commendable" and recommended that construction be abandoned permanently.[76] BC Rail was deeply in debt: an internal audit revealed it would be paying $42 million annually by 1980 just to service its debt.[77]

Prior to the official abandonment (perhaps part of the motivation for abandonment in the first place), contractors began to demand compensation for lost revenue and their own over-expenditure. Private-sector grievances were made public in the aftermath of the provincial government disclosure of new BC Rail deficits in June 1976. Revenue issues were compounded when many of the projected customers, including coal mines and pulp mills, did not materialize. Later in 1976, MEL Paving, a Red Deer contractor hired to complete the last section of the extension between the Stikine River and Dease Lake, sued BC Rail, alleging it had been strategically misled about the scale and complexity of the work. MEL sued for fraud, claiming a wilful discrepancy on the part of BC Rail between earthwork estimates and actual tonnage to be moved. The two parties settled out of court before the Supreme Court of British Columbia finished hearing the case. Environmental organizations tried to secure "remedial action" and to force BC Rail to properly put the extension to bed. In spite of these efforts, there was little remediation or securing of vulnerable slopes: the extension remains much as it was when it was abandoned, the only maintenance completed by resource companies at particularly troublesome spots in order to make the road passable for their trucks, equipment, and personnel.

The Stikine Bridge, built at the juncture where the extension crosses the river to continue its path north to Dease Lake, has become the signal icon of the extension debacle for many locals. Completed in 1977, at a cost of $3 million, it was promptly closed and blocked with fencing, upon orders from BC Rail management. This closure occurred only months after its completion. To compound the effect of incompetence, the Stikine Bridge would have been submerged several hundred metres underwater if BC Hydro had gone ahead with its concurrent plans to build a series of dams upriver on the Stikine, a prospect outlined in the next chapter. The Stikine Bridge stands today, a talisman of the lack of planning, integrated management, and forethought of BC Rail and its Dease Lake Extension.

The reasons for abandonment are complex, but the unbuilt environment here has produced side effects that connect the abandoned extension to problems of increased area access, the contemporary resource boom, and questions around resource politics. The northern dream of a railway through the Stikine did not die with the abandonment of the Dease Lake Extension in 1977. In October 2005, a consortium of investors, lobbyists, and politicians who had largely forgotten about the profound effects of earth-moving, culvert construction, and increased access in the Upper Skeena, the Klappan, and the Stikine, announced a feasibility study to examine various transportation routes between Fairbanks and the BC Rail system. This new study group was going over well-travelled ground in Alaska: the rail link had long been a goal of public officials and private investors in the state.[78] Making British Columbia a hub of continental integration had also been W.A.C. Bennett's dream. He wanted the extension to connect British Columbia with Alaska. From Dease Lake it would eventually push north to Watson Lake and then west towards Anchorage. By 2005, it was the view of enthusiastic Alaska Canada Rail Link boosters that the Stikine would once again be a place through which people, things, and ideas moved. But unlike Bennett's galvanizing political style, an Alaska Canada Rail Link group newsletter emphasized the legacy of scientific and technological data produced by its commissioned studies: "If the study reveals that the railway is not financially feasible at this time, the extensive body of knowledge that has been developed will become a legacy for long range transportation planning in the north and the criteria for future rail investment viability will be determined."[79] The boosters warned that development would be costly and risky: "This line segment [requires] very high capital, maintenance and operating costs. This is a high energy consumption line and the right-of-way would have a high risk of exposure to natural disasters."[80] But these concerns did not seem to matter: the "working scenario" proposed to the governments of Alaska and the Yukon incorporates large sections of the abandoned Dease Lake Extension as the material basis of the proposed rail link between Alaska and southern British Columbia.[81]

The Alaska Canada Rail Link scenario brings us back to Fortune Minerals' plans to move coal by rail. Indeed, there is a great deal of overlap between Fortune's proposed route, outlined in the introduction to this chapter, and the vision developed in the Alaska Canada Rail Link scheme. Fortune even used the Alaska Canada Rail Link proposal to buttress its rail plans. The necessary upgrades – additional ballast and steel rails, and bridge and earthwork – would be expensive, but the expectation was that other potential users could share some of the cost. If the extension were activated again, Fortune could transport materials by rail to Prince George, where ore could be loaded onto trucks to be taken via the highway system to the Ridley Coal Terminal at Prince Rupert to be shipped overseas, most likely to China. Within this frame, the abandonment of the Dease Lake Extension has another, unforeseen potential environmental legacy – as a conduit for the export of coal. It stands as a failed infrastructure project, with effects that continue to percolate long after its demise. After a period of sharp public interest and concern by the press, the extension seems to have faded from public environmental consciousness. It was abandoned both as an integral component of transportation infrastructure and as an environmental concern. Yet many of the environmental and social dislocations associated with the extension have metastasized over the years of neglect. The rail bed, still in the "wrong location," still eroding into the Klappan River, and still barely holding up over collapsed culverts, presents a cheaper transportation option – a kind of literal and metaphorical "shortcut" – for Fortune and the new northern-rail dreamers in Alaska.

An epilogue to the Dease Lake Extension story reinforces the impression of the short historical memory of railway and mining pioneers in northwest British Columbia. Facing off against weak global metallurgical coal prices and renewed Tahltan opposition to its mountaintop removal plans in the Klappan, Fortune and its new partner sold their interests in the project in

May 2015 for $18 million.[82] Given the uncertainty over the coal industry, and over the potential transportation options, the sale was probably good business practice, and it allowed Fortune to focus its extractive attention on its Nico property 160 kilometres north of Yellowknife. But the identity of the buyer should raise the reader's eyebrows. The province brokered an agreement whereby BC Rail bought the sixty-one contiguous coal licences that comprised the Klappan tenure, giving the final act an almost teleological edge. The agreement also requires BC Rail to perform and pay for any road reclamation and repairs. Almost forty years after the decision to terminate construction, BC Rail is once again responsible for the leftovers of its abandoned Dease Lake Extension.

3
Corporate Ecology: BC Hydro, Failure, and the Stikine-Iskut Project

In the latter half of the twentieth century, a growing awareness of the negative social and ecological effects of large dams prompted the emergence of an international anti-dam environmental movement. In British Columbia, this emerging environmentalism faced off against the pro-development stance of Premier W.A.C. Bennett. At the crux of Bennett's attempts to lure development and investment to the province was a comprehensive plan to harness the hydroelectric potential of British Columbia's northern rivers. Massive dams laid across great rivers would harness the kinetic energy of accumulated water to produce electricity that could then be carried across the province in transmission lines to power a new and prosperous industrial future for the province and its people. To this end, BC Hydro surveyed and plotted the "power possibilities" of the provincial river network, with a particular focus on the "unused" rivers of the north.[1] Dozens of northern rivers were measured and itemized – water turbidity, temperature, head, and flow became new scientific metrics for understanding the value of rivers. In the late 1970s, BC Hydro announced plans to dam the Stikine and Iskut Rivers. In the utility company's view, these dams would extend a trajectory of development already well established in the province, one that had already seen the construction of dams on the Peace River, agreements with Americans about dams on the Columbia, and concerted but ultimately frustrated efforts to dam the salmon-bearing Fraser.

These postwar decades were the age of the dam, not only in Bennett's vision and in BC but all across Canada and North America. The great surge of megaproject construction in Canada saw engineers redirect the flow of the Nechako; reorient the hydrology of Manitoba with the Churchill River diversion scheme; reengineer the drainage of Quebec and Labrador with the Manic, Churchill Falls and James Bay projects; and modify the St. Lawrence into a reliable commercial marine thoroughfare. Foreshadowed by American projects like the Tennessee Valley Authority and the Grand Coulee Dam, and mirrored in Russia, China, and elsewhere, these dam projects were the most celebrated Canadian expressions of what anthropologist James Scott called "high modernism": faith in science and technology, in the simplification and careful management of the natural world, and the conviction that the mastery of nature and the expansion of industry would result in progressive social and economic prosperity.[2]

But in Canada, and particularly in British Columbia, the course of river development did not always run smooth, complicating the logic that large-scale interventions in the environment would bring prosperity across the board. In a study of the abandonment of damming projects on the Fraser River, Matthew Evenden has suggested that the impetus behind dam construction was the desire to implant "new technologies of power," wherein science and technology fostered progress and profit.[3] Tina Loo has suggested that the damming of the Peace reflected the politics of scientific engagement under high modernism, arguing that the self-declared "progressive" philosophies behind the super-developments of the 1960s actually resulted in massive social and environmental dislocation.[4] In Quebec, Cree interactions with Hydro-Québec and its development of James Bay's hydroelectric potential have allowed Caroline Desbiens to show the emergence of a new geography of resource exploitation and national identities.[5] For both Cree and Quebecois, shifting ideas about nature, nation, and gender crystallized around threatened landscapes, livelihoods, and lifeways. New work on megaprojects seeks to understand the impact of massive, often state-led, infrastructure and energy projects, as well as how modern technologies and scientific engagements remake environments.[6]

In this chapter, I examine the planning, management, and debate surrounding BC Hydro's power scheme to generate on the Stikine and Iskut Rivers. The environments envisioned by dam proponents and detractors show how the battle over nature in the Stikine galvanized interest in the area, promoted conflicting development possibilities, and used the region's history to mobilize support for different viewpoints.

The Stikine-Iskut Project

Between 1973 and 1983, BC Hydro tried to develop the Stikine district with a massive power venture by damming the Stikine and Iskut Rivers. None of the dams were built, as construction plans ran into opposition from local and metropolitan groups. The ensuing debates over resource use hinged on the ways various stakeholders saw their relationship to the plateau. Interest in the hydroelectric potential of the region dates to 1961, when the BC Lands Service first suggested the possibility of adding the Stikine-Iskut basin to the province's hydroelectricity reserve. In 1964, the BC Water Investigation Board identified potential dam sites, which led to an order-in-council that restricted mining activity and held lands near these sites in reserve. Small-scale surveys continued over the next decade. In the early 1970s, vacationing *Vancouver Sun* reporter Bruce Larsen stumbled onto a dozen men engaged in survey and core drilling work for Brinco, builders of the massive Churchill Falls hydroelectric system in Labrador. Larsen guessed that Brinco was planning a dam, employing the following logic, which, in retrospect, connects dams to development projects:

> 1) the provincial government is rushing completion of the Stewart-Cassiar road, 2) the provincial government will have completed the PGE railroad [Dease Lake Extension] south of the Stikine and into Dease Lake by 1974, 3) at least 34 mining properties are ready to go into production as soon as the railroad arrives, 4) the area right now is swarming with helicopters under contract to mining companies, 5) there are more geologists in the hills than goats, 6) mines need hydro power.[7]

Brinco initially denied interest in the Stikine, probably because it had no permit to work there and had not submitted a formal proposal to survey the area. Brinco later admitted its interest and engaged in a protracted public relations exercise, which continued until 1977, when it submitted a formal proposal.[8]

Meanwhile, BC Hydro commissioned its own feasibility study in 1977 (published the following year). The study "found that, based on preliminary data, the Stikine-Iskut development was technically feasible and economically attractive."[9] BC Hydro developed a plan that focused on damming five sites (Figure 3.1).

Two dam sites were selected by BC Hydro on the Stikine River: Site Zed, suitable for a 270-metre (75-storey) archway at the head of the Grand

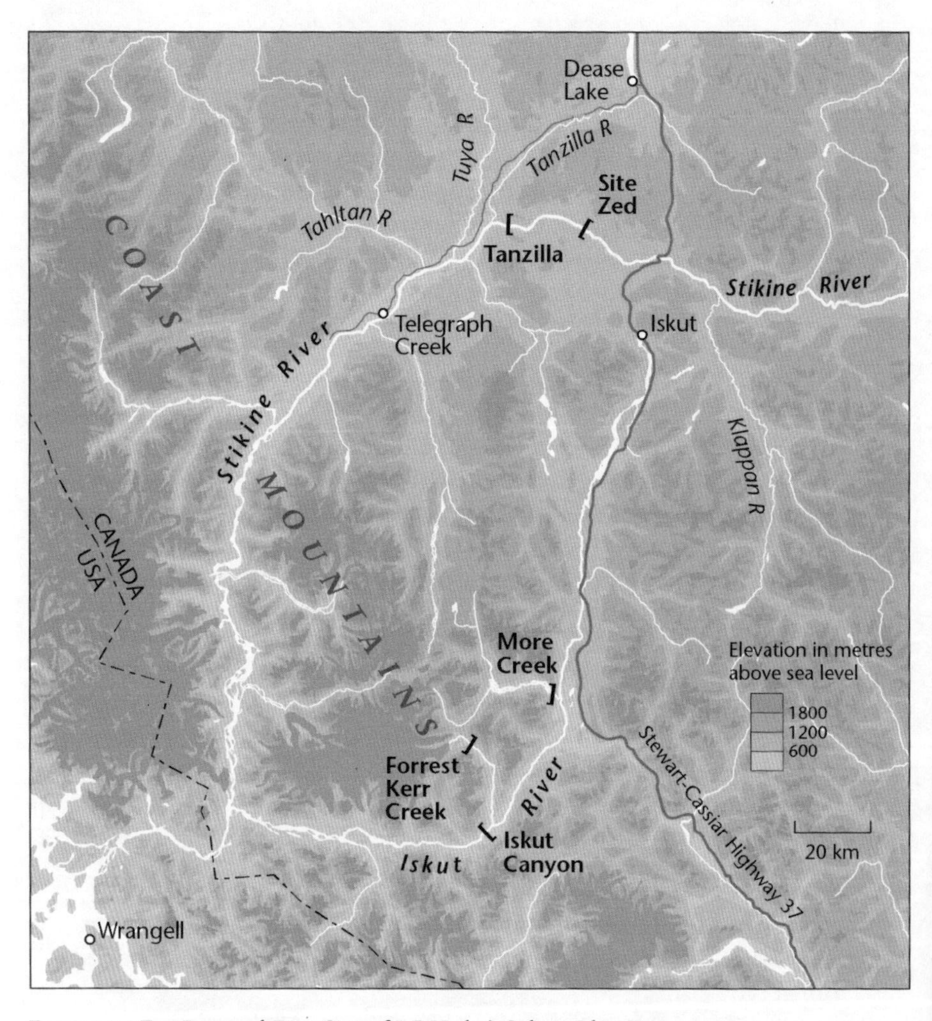

FIGURE 3.1 Five Proposed Dam Sites of BC Hydro's Stikine-Iskut Project. Map by Eric Leinberger

Canyon of the Stikine; and Tanzilla, thirty-seven kilometres downstream, where a 193-metre structure would be built at the confluence of the Tanzilla and Stikine Rivers (see Figure 3.2 for a visual representation of the landscape effects of these proposed structures). Behind these projects, two massive proposed reservoirs, with surface areas of 12700 hectares and 900 hectares, respectively, would have flooded the Grand Canyon of the Stikine, an extraordinary tectonic marvel fifty-four kilometres in length with walls close to one thousand metres in height in some places.[10] Three smaller – though no less imposing – structures were planned for the Iskut River, the Stikine's largest tributary and the primary site of proposed mining development. Two dams, a 158-metre monolith in the Iskut Canyon and a 135-metre structure at More Creek, and a smaller diversion dam at Forrest Kerr Creek, were to create two reservoirs, twenty-five and thirty kilometres long, providing roughly one-third of the Stikine-Iskut project's anticipated 2800 megawatts of power – a spectacular amount considering that BC Hydro's thirty-two other generating facilities, combined, produced 7500 megawatts of power in 1978. The cost of the proposed project was equally staggering: an estimated $7.6 billion, the project would have been the biggest capital project ever undertaken by BC Hydro.[11] With such massive investment on the line, BC Hydro commissioned reams of data to support and justify the project.

BC Hydro's effort was complicated by a number of factors. In 1980, the BC government amended the BC *Utilities Commission Act,* creating a new regulatory body with formal oversight over power development, power distribution, and power sales in the province.[12] BC Hydro set its own regulatory terms before the implementation of the *Utilities Commission Act:* the regulatory reform of the British Columbia Utilities Commission ensured that the utility was no longer autonomous in decisions regarding energy planning. In addition, the institutionalization of an environmental impact assessment apparatus was a central component of the *Utilities Commission Act.* This was a major departure from previous management, assessment, and regulation of energy resources in the province. In the coming years, the construction of new generating facilities would take a back seat to the economical operation of existing ones.[13]

Northern development goals faced the formidable but often ambiguous environmental politics of the sixties and seventies in British Columbia. It took many forms but wilderness protection was front and centre of the environmentalist agenda.[14] From this conjuncture sprang two influential environmental organizations dedicated to preserving the Stikine in its "natural" state – Friends of the Stikine, based in North Vancouver, and

Residents for a Free-Flowing Stikine, based in Telegraph Creek, the only permanent settlement on the river. The Tahltan, politically active and increasingly involved in large-scale business disputes within their territory, presented a consistent chorus of dissent, primarily in the language of unresolved land claims but also appealing to the ecological and cultural integrity of the watershed. Other interest groups also demanded more information and input, questioning BC Hydro's transparency and data on issues ranging from power consumption projections to fisheries damage to international legality. A cat-and-mouse game of environmentalist agitation, cagey corporate engagement, and Tahltan frustration continued until the fall of 1983, when BC Hydro, citing mounting debt, a power surplus, and tentative future consumption prospects, reluctantly announced the postponement of the in-service date of the first installations – from the early 1990s until the twenty-first century. In 2000, the postponement became permanent when the Tahltan negotiated a land and resource management plan (LRMP), formally protecting 25.4 percent of the Stikine watershed, including the Grand Canyon of the Stikine, from major development.[15]

Although the dams never materialized, the discourse and debate around them had very real material consequences for the landscape and the way people interacted with the environment. Through these discourses, people came to know (and think about) the river differently, and this changed the materiality of the watershed. People knew the river differently partially as a function of the creation and codification of knowledge about the area. There was little institutional knowledge about the Stikine watershed – this is repeated and emphasized in every report commissioned by BC Hydro. This created an archival "blank slate," with knowledge gaps that had to be filled. The constant refrain is that BC Hydro and its contractors (and by extension other official bodies) knew very little about the area; therefore, "scientific" studies were necessary to correctly identify the inherent values of the landscape and the value that should be ascribed to them. In the language of BC Hydro, this was to ensure both proper use and proper protection. The "environment" came to be organized into a rational typology in which contingent uses were circumscribed.

"Technically Feasible and Economically Attractive": Framing the Constituents of the Debate

An essential part of the planning process for major new projects is the early identification and assessment of potential project effects on wildlife, forests,

> agriculture, recreation, archaeology, and human settlements. BC Hydro recognizes the need to avoid unfavourable impacts wherever possible, and the responsibility to mitigate or compensate for them when such impacts cannot be avoided. To obtain the data necessary to make informed judgments, BC Hydro's planning process provides the opportunity for exchange of information with environmental specialists, appropriate government agencies, and the public at large.[16]

BC Hydro's interventions into the Stikine were framed by a nascent environmental prerogative that was gaining momentum in Canada. Environmental impact assessment procedures were established at the federal level in Canada in late 1973 (as guidelines, though concrete formal protocols were implemented in the 1990s), and were buttressed in fits and starts by various local assessment cases throughout the rest of the decade. The development of environmental impact assessment protocols followed growing concerns over the scale of environmental impact and potential social dislocation that plagued the aftermath of the megaprojects that had increasingly defined resource development by the 1970s. Commentators have highlighted the importance of the Mackenzie Valley pipeline inquiry (1974–77) and the James Bay and Northern Quebec Agreement (1975) in moving the federal environmental assessment and review process (EARP) forward, alongside Indigenous claims-based assessment procedures.[17] BC Hydro developed plans for the Stikine-Iskut project within this context of a nascent assessment culture.

Between 1977 and 1983, BC Hydro commissioned eighteen studies in the areas of forestry, fisheries, archaeology, caribou migration, climate and meteorology, economic geology, river regime and morphology, furbearers, land use, mountain goats, recreation and aesthetics, resource evaluation, socioeconomic considerations, the Stikine estuary, townsite planning, vegetation mapping, transmission, and wildlife generally. The scope and range of these reports are difficult to summarize (see Appendix for details). These specific studies were complemented by a series of intermittent progress reports that synthesized new developments and data associated with the project and provided a general context for assessment. BC Hydro did some of these studies in house, but most were handled by outside contractors, primarily based in Vancouver or other major western cities. It is possible to identify patterns and recurring themes in these reports, but I am concerned here with how these reports produced knowledge about the resources of the Stikine and the impacts that the dams would have on those resources and the communities that relied upon them. More importantly, the reports functioned as frames for analysis and debate about the dams themselves. They plotted the terms in which various interests were required to speak about the river and its reorganization. As BC Hydro's environmental studies coordinator Joe Alesi put it, "We're looking at the real issues, not every dicky-bird and squirrel."[18] Or, in more conventional contracting language, the assessments were needed "to provide the follow-through, and to connect the loop between doing the environmental studies and preparing the environmental impact assessments and actually carrying out specific environmental protection functions on projects."[19]

Notions of expertise and calculation gained real traction through the creation and dissemination of these reports.[20] Bureaucratic and institutional power pushed knowledge to support BC Hydro's designs. The framing of environmental and economic concerns allowed BC Hydro to demarcate how opponents and interlocutors could engage with development. If it is axiomatic that knowledge and power are intimately interconnected, then the production of geographic knowledge in these reports obscured previous forms of knowledge that inhabitants had used to organize their interactions with the river.[21] The reports commissioned by BC Hydro's scientific program often obscured as much as they generated. The production of non-knowledge and fragmentation contributed to BC Hydro's development aims by marginalizing or co-opting critique. "Non-knowledge" in this sense refers to how the data produced and the methods used by BC Hydro were largely unavailable to the public at the time. The creation of new scientific data drew boundaries around how interested parties could talk about nature in the Stikine, even as it obscured the tangible scientific terms on which BC Hydro's plans were based.[22] There was little "consultation"; meetings orchestrated by BC Hydro controlled the terms of the debate and the release of information. In the Stikine, the creation of non-knowledge was integral to limiting resistance and to managing discourse on damming. This issue has been exacerbated over time as corporate records and consulting reports have been moved to provincial archives and libraries. Arguably, the production of non-knowledge can be viewed as an effect of the social world of archives, a powerful bureaucratic practice of making meaning in both history and the present.

A series of reports on transmission line routes illustrates the argument about knowledge. The environmental impact assessment section in the 1982 report draws parameters of possibility around the study, by claiming that the study was hamstrung at the outset by factors outside its control, outside the realm of controllable scientific variables: "The preliminary environmental and social impact assessment studies for the Stikine-Iskut transmission project are part of the most complex transmission studies ever attempted in British Columbia. The vastness of the study area and the general lack of detailed information added to the difficulty of carrying out the environmental studies."[23] Here, complexity and size are given as explanations for the lack of knowledge and expertise. Science is hampered by a lack of modernity, spatial marginality, and obstructed mobility; technology cannot transcend the spatial and temporal dislocation. The predetermined difficulties and engineering ambitions absolve BC Hydro of the inevitable ecological mistakes through a kind of pre-emptive *mea culpa*.

Some of the disruption is glossed over by the rational organization of information into distinct categories. BC Hydro categorized the created and collected information because it could not envision any other way of engaging with it or presenting it: "After consolidating available information and collecting new data in the field, each environmental study team member assessed the impact of transmission line construction along proposed corridors from the point of view of his/her individual, specialized subject matter, e.g. forestry or wildlife," or fisheries or geomorphology.[24] Ecosystems are abstracted into categories, distinct from one another to ease planning. But the result is inconsistent with most basic ecological knowledge.[25] It hardly needs to be mentioned that the information that BC Hydro sought to collect and collate is scientific, measureable knowledge, not Tahltan knowledge garnered through centuries of experience. James Scott describes this process of the rendering of abstract ecological phenomena into measureable, contingent values as the enactment of "legibility."[26] Once environments are made legible to the state and its administrative apparatuses, they can be ascribed value and rated according to their desirability and worth.

After field studies were completed, the BC Hydro study team met for several days in order to compare observations, determine the relative importance of resources within each corridor, and assess the combined resource consequences of transmission line construction within the corridor. The consultants converted their individual results into a common rating system with eleven categories – ranging from very substantial benefit to severely negative impact. The ratings categories are used throughout the source environmental report, in the consultants' basic descriptions and their comparisons of the impacts.[27]

Animals and habitats were subject to calculation and categorization, where the "relative importance" of caribou was somehow measured against

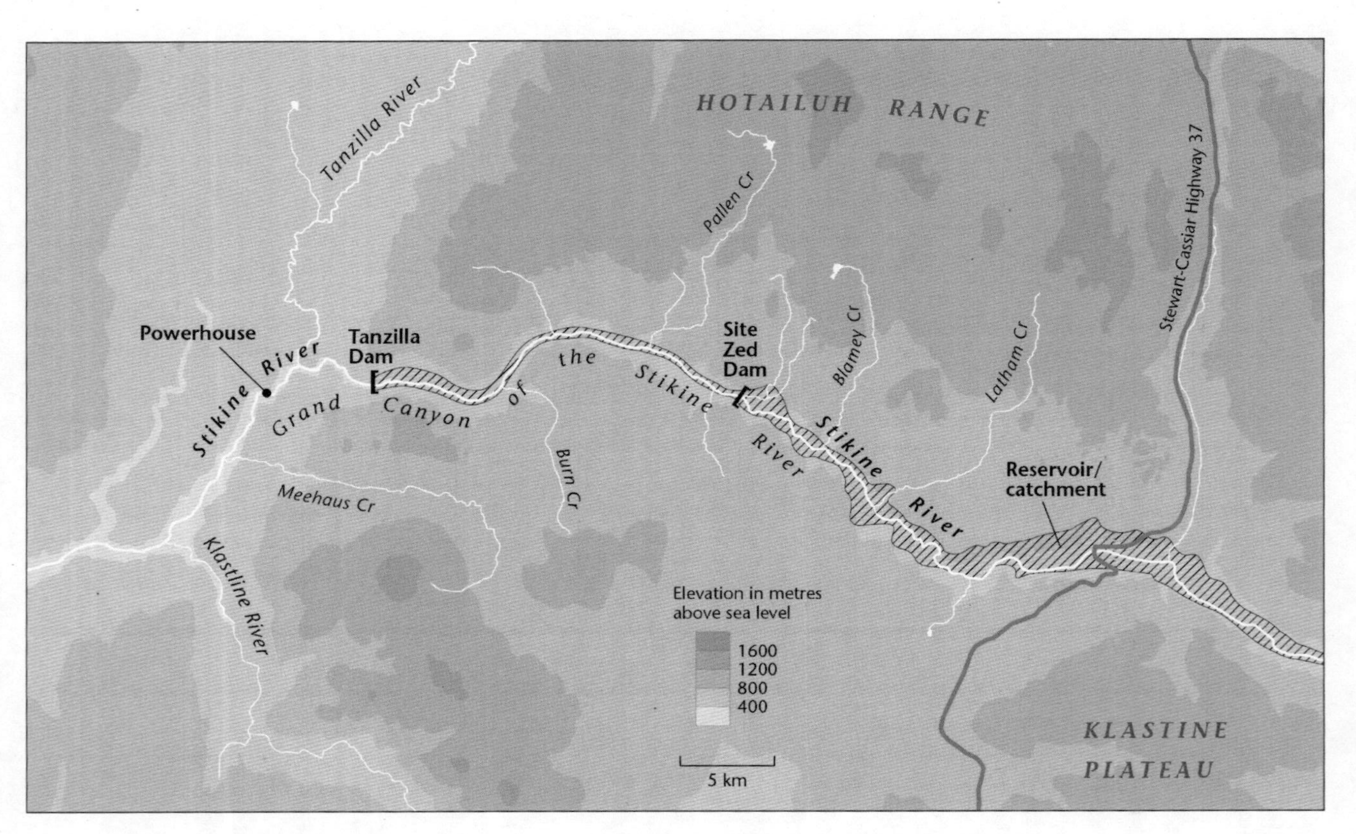

FIGURE 3.2 Two Proposed Dams and Reservoirs in the Grand Canyon of the Stikine. Map by Eric Leinberger

that of other objects of analysis – moose, or a stand of spruce, or the possibility of finding an ancient piece of obsidian jewellery buried in the earth of an abandoned fishing station.

A more pointed example is a report commissioned by BC Hydro on the mountain goat population of the Stikine Canyon. The authors sought to map out the "effect of exploration, proposed construction and post-development activities (including flooding) on a mountain goat population residing within the Grand Canyon of the Stikine. Preliminary evaluation suggested that partial flooding of this habitat and increased public access could impact resident mountain goats." These impacts included "localized goat mortality" and "possible temporary range abandonment."[28] The consultants estimated that 56 of the estimated 316 mountain goats living in the canyon would be killed when the Grand Canyon of the Stikine was flooded after the construction of a 75-storey dam (Figure 3.2). There is no doubt that fifty-six dead goats is a somewhat arbitrary outcome of some consulting arithmetic – or, to borrow novelist Michael Chabon's apt phrase, an outcome of "feats of inspired guessing"[29] – but a more relevant question might be, how is 56 of 316 understood to be an acceptable number? If so, for whom is it acceptable? For BC Hydro, acceptability was largely left for others to judge. After all, this was an information-gathering exercise; wilderness values, understood in the context of the disconnected, diplomatic, and bureaucratic language of BC Hydro's commissioned data, were extraneous to the conversation. Problems also derived from what was included in and excluded from the study. For example, the canyon is home to a variety of raptors, including the endangered peregrine falcon, but there was no mention of the falcon in the report, or in any other reports commissioned by BC Hydro. The falcon was further threatened by the fact that it was not included in BC Hydro's terms of reference.

BC Hydro studies were narrowly conceived, intended to focus on limited topics. Most were intended to establish "baseline data," as starting points for the analysis of environmental disturbance. The reports were intended to serve as references to a fixed point in the past, anchoring the nature of the Stikine Canyon as it faced change in the future and BC Hydro determined which other possible resources might be endangered or lost. The economic geology report was designed to "describe and evaluate mineral, rock, oil and natural gas deposits that would be affected by the proposed hydroelectric development on these resources also described." This study included an "inventory" and an "assessment," that is, a cataloguing of resources for an accounting and comparison of their best use.[30] Reports such as these had far reaching consequences for mineral and energy development in the region. Here, potentially "lost or foregone" resources were afforded real economic value, even while still in the ground, so that the cost of not exploiting them could be measured against the benefits of other planned uses.

The river regime and morphology study mobilized exclusivity in much the same way. The study was motivated by two goals, both precursors to future work but not scientifically useful in and of themselves: first, "to provide information for use in the project feasibility study and secondly, to provide information for the environmental impact analysis." The authors then provided a comprehensive list of study proscriptions and project entitlements to be used in the assessment of flood levels, flood frequency, river flow regime changes, ice observations, suspended-sediment sampling, and water temperature gauging. The report concludes that dam construction would cause summer river flows to decrease and winter flows to increase; there would be more siltation and less seasonal temperature variation. Effects on fish and aquatic vegetation were not discussed. Fish and aquatic vegetation were under the purview of another group of researchers.[31] The reports may have been designed to facilitate future research endeavours, but none of those commissioned by BC Hydro had provisions for long-term study or for the continuation of studies beyond the reported field season. This was partially because of the provisional nature of BC Hydro's dam plans, but it also reflected the arbitrary and exclusive nature of the technical survey work undertaken in support of the damming project.

The discursive powers of absence also emerge from the archaeological fieldwork. Aresco of Calgary was contracted to set up an "evaluative device" and to ascertain the (wonderfully contradictory) "heritage potential" of the Stikine watershed. Using a variety of archival, photographic, and visual survey techniques, Aresco determined that there would likely be some impact on undiscovered archaeological sites, especially at the confluences of rivers where people had congregated. Impacts at these sites would be minor, however, because it was anticipated that little archaeological material would be located there. Contradicting of Tahltan ethnographic material and oral histories, Aresco consultants claimed that the Tahltan were historically more dependent on large mammals than on fish and did not use the river much, for fishing or transport, until the advent of the fur trade. In this case, the sites and histories of Tahltan use of animals and fish were mobilized by BC Hydro and its consultants to push for alternative human contemporary uses of the environment. Most of the archaeological sites identified in a twelve-day field season were on the plateau above the proposed reservoir site and therefore unlikely to be affected by more than incidental impacts from transportation and infrastructure. Furthermore, the report cites ethnographic data gathered in the analytical trope of early twentieth century "salvage ethnography" as the definitive evidence on Tahltan hunting grounds and culture. In this case, knowledge (or the lack of a site for knowledge or evidence) operated to dispossess the Tahltan and helped to justify the damming project.[32] BC Hydro's commissioned reports were predicated on a system that gathered information according to set procedures but often overlooked Tahltan knowledge in the process.

The discursive and ideological effects of BC Hydro's knowledge creation notwithstanding, the work of contractors and technicians produced actual material effects on the landscape. By December 1981, BC Hydro's "investigations" at Site Zed had amounted to "39 diamond drill holes, totalling 6500m; 3 adits totalling 1075m; seismic refraction and gravity surveys; auger holes and test pits for materials exploration."[33] Though they were most extensive at Site Zed, explorations of this type occurred at every planned site. Further investigations at Site Zed would require a very heavy, track-mounted Becker hammer drill to punch through the overburden material. The drill would have to be flown in to the area, which would necessitate the construction of a 1600-metre airstrip to accommodate the Hercules aircraft transport, plus a seven-kilometre road to the site itself. Alternatively, at a savings of $800,000, BC Hydro proposed to build a forty-six-kilometre track from Highway 37 to Site Zed.[34] None of this infrastructure was ever built, but issues of access were important, and not just for BC Hydro employees and contractors. "Indirect disturbance" would occur due to increased hunting opportunities and private exploration access. More directly, BC Hydro admitted that its investigations themselves could have specific consequences: "1. Damage to soils and vegetation from excavation, drilling and placement of equipment; 2. Disruption of wildlife and a temporary reduction of wildlife habitats; 3. Disruption of trapping and opportunities for guiding; and 4. Possible disturbance of archeological sites."[35] In BC Hydro's assessment, these were direct but localized impacts, unlikely to galvanize opposition outside of that from interested locals and small-scale metropolitan niche organizations.[36] But BC Hydro may have underestimated the level of protest and animosity encouraged by its tactics and investigations.

"Intervenors": Environmentalist Critiques

Approximately thirty groups, termed "intervenors" by BC Hydro, actively protested against Stikine-Iskut project. Friends of the Stikine and Residents for a Free-Flowing Stikine were the most prominent among them.[37] Formed in Vancouver in 1980 at a meeting sponsored by the Sierra Club of Canada, both Stikine groups showed a willingness to work within mainstream politics. They sought to educate and influence public opinion, but they engaged in little direct action.[38] They embraced a reformist strategy that focused political energy on the publication of newsletters, participation in limited public debate, and media coverage.

Based in Vancouver and connected to various communities and members of local universities, Friends of the Stikine was a more publicly prominent voice in British Columbia. Its goals were simple: "to enhance public awareness of the advantages to be gained through preserving the Stikine River and its immediate environs in its natural state" and "to develop ... alternative policies for managing it, in keeping with the wilderness character and the aspirations of native people."[39] The Friends of the Stikine campaign focused on "effects" and "impacts" in much the same way as Hydro's reports. The organization's first newsletter justified opposition to the dams by listing the "primary effects" and "secondary effects" of dam construction: the destruction of the Grand Canyon of the Stikine; the negative social impact on the area's small communities and on Indigenous lifeways; the irreversible commitment of the area to energy production; the abrogation of unsettled treaty rights; the uncertain impacts on wildlife, including salmon populations; the increased access for hunters and mineral prospectors and what that might mean; and the ecological disturbance of the Stikine delta and deterioration of the fishery.[40] This was the basis of an argument against the construction of the dams. But without the scientific and technical evidence of the kind that BC Hydro was creating (but not disseminating), groups like Friends of the Stikine were unable to motivate widespread support for their cause.

Indeed, the language of "impacts" was used by BC Hydro representatives to analyze and disarm opposition. Bentley Le Baron, a contractor with Canadian Resourcecon, was tasked by BC Hydro with detailing the social and economic impacts of the prospective dams. He grouped all opposition together by suggesting that "wilderness preservation" was the central organizing principle of opposition efforts.[41] By framing wilderness as the primary focus of the opposition, the report was able to reduce the complex interests of environmental groups to something relatively simple. The appeal to wilderness also presented the Stikine as a wild space, a space removed from human interaction. This version of opposition discourse meshed passively with BC Hydro's corporate discourse, and may have allowed BC Hydro significant leeway in incorporating and co-opting the concerns of a diffuse and often antagonistic resistance movement.

Residents for a Free-Flowing Stikine drew its main support from people who lived in the watershed. Members of this group were more willing than the Friends of the Stikine to use non-mainstream resistance tactics and showed greater commitment to local political concerns. In this sense, their opposition to the dams was more complex. Writing in the *Northern Times* in September 1980, spokesperson Joe Murphy predicted devastated communities as he and other Telegraph Creek locals decried southern administrators who made unilateral decisions about their land and livelihoods. Group members envisioned development based on

> a just settlement for the land claims of the Tahltan People, the original inhabitants of the country to be flooded. We see the development of a regional economy based on commercial fishing, agriculture and the older ways of trapping and outfitting. We see the preservation of ... the Grand Canyon of the Stikine. We ask for nothing more and nothing less than the right, as local residents, to have some say about how future developments will best serve the needs of our own people and region rather than serving the needs of corporations across the border. To BC Hydro we must say: "We have too much to lose – there can be no compromise. We must stop the dams."[42]

Sensitivity to local economic concerns, particularly fisheries, and a willingness to promote the resolution of outstanding land claims were hallmarks of the campaign by Residents for a Free-Flowing Stikine. In these, they had an ally in the Anglican Church of Canada, the church group with the longest-standing institutional presence in Telegraph Creek. These groups were early proponents of the notion that the Stikine Canyon should be turned into a park.[43] There was mixed success in partnering with the Tahltan, who often regarded environmental groups as interlopers in much the same way as they perceived BC Hydro and other resource companies and their employees. There was particular ambivalence about the perceived external influence of the Friends of the Stikine, who were essential in bringing the Stikine into the consciousness of people outside the region but who were also perceived as outsiders.

Efforts of environmental groups were hampered by BC Hydro's policy regarding the data it was collecting. No information would be made available to the public while studies were ongoing. The interim reports that were released by BC Hydro were removed from their ecological context and from the circumstances of their creation. Wildlife biologist Rosamund Pojar, a charter member of the Friends, summed up the frustration of advocacy within an analytical vacuum. Writing in the *Telkwa Foundation Newsletter*, she condemned Hydro's "policy of conducting fragmented environmental studies to obscure its grand plans."[44] For Pojar and her colleagues, the information imbalance handcuffed public interests and created a lamentable situation in which "predictions about the impact of the dams based on information available to public agencies will necessarily be crude and speculative."[45] Indeed, this fit with Hydro's mandate of the creation of non-knowledge in these scientific studies. Because little was known about the biophysical qualities of the Stikine watershed, and even less about the actual possible effects of the dams, opposition was limited to a series of provisional arguments.

"The River is Central to Our Lives": Local Actions and Reactions

The Tahltan's social and economic livelihoods have been based on area resources for thousands of years. The river was the centre of this northern lifeworld: it was the primary transportation conduit, the main food basket, and a prominent cultural symbol. The dams threatened to challenge that social and economic way of life and worldview by limiting the river's function and meaning. Alongside the dam debates and BC Hydro reports, there remained unresolved questions about the health of the fishery, the status of important cultural sites, employment, and the hunting/guiding industry, among others.

A thin consensus emerged among Tahltan on BC Hydro potential. Tahltan leadership, particularly within the Telegraph Creek band, was vocal from the outset. Spokesman Gordon Franke, present at the January 1980 meeting in Vancouver that led to the formation of the two Stikine-focused environmental groups, said that "if BC Hydro puts a dam on the Stikine, it will exterminate the Tahltan people because the river is central to our lives."[46] All of the area to be flooded fell within the 130,000 square kilometres of territory claimed by the Tahltan. Tahltan leadership invoked the 1910 Tahltan Declaration, which proclaimed, "We claim the sovereign right, to all the country of our tribe – this country of ours which we have held intact from the encroachment of other tribes, from time immemorial, at the cost of our own blood ... because our lives depended on our country."[47] BC Hydro's dam project threatened that sovereign right, even though that right remained unresolved in Canadian law. The Association of United Tahltans (AUT) was formed in 1975 to represent the Telegraph Creek and Iskut bands in land claims negotiations; the association spoke for the Tahltan in dealings with BC Hydro. Association president George Asp made it clear that the Tahltan "aren't going into our aboriginal rights claims negotiations after the fact of these dams. It's a matter of survival for us."[48] But the knowledge possessed by AUT members, accrued over years on the land, proved inadequate in negotiations about the project. Non-knowledge could be produced when established forms of knowing were superseded by new ways of knowing; in this instance, the Tahltan were denied access to the scientific and technical knowledge that BC Hydro was organizing and BC Hydro was unwilling to share the data it had garnered from its fieldwork. Chief band councillor Ivon Quock recalled a BC Hydro public affairs official visiting Telegraph Creek in 1980. The official had copies of BC Hydro's environmental studies, "but he only kept them here for a number of hours, and he told us there weren't enough copies to leave any behind. He took a lot of notes, and then he went back to Vancouver and told the papers the Indians were in favour of the dams."[49] The Tahltan and other opposition groups complained repeatedly about the lack of transparency in BC Hydro's process and their inability to access information through BC Hydro systems.

To plug the gap, the AUT initiated its own environmental assessment and inventory study. The Stikine Basin Resource Analysis was envisioned as a three-and-a-half-year project with a proposed budget of $7 million. It was never completed, due in equal parts to the difficulty of large-scale fundraising and the negligible value of the analysis once dam construction had been deferred. But the study was instigated because the AUT felt it had no choice if it were to protect Tahltan land claims: "Why should we have to duplicate the research Hydro is doing?" Asp said. "We don't want to take an adversarial approach from the outset. But until we compile all the needed information, we have no choice but to oppose those dams."[50] The Tahltan undertook a systematic inventory of valued resources in the watershed, in effect replicating the data collected by BC Hydro technicians and consultants. Believing that "things are about to change," the Tahltan aimed to "realistically assess the overall value of this unique area of Canada to ourselves as native people and to further assess the impact of [BC

Hydro's] proposed developments on the opportunities that the region holds for us in the future."[51] While the methods used and the data created were similar to those used and collected by BC Hydro, the objective of the AUT was very different.

The AUT was particularly concerned with assessing opportunities for future watershed uses that would be imperiled by dam construction and the associated environmental changes. The goal was an impact assessment that focused on the evaluation of natural changes over a simple description of predicted effects. The scale of analysis was also longer and included reference to other resource activities: "The initial objective will be to develop an adequate data base in order to assess the effects of the hydroelectric development in a meaningful way. Ultimately, spin-off benefits will include use of the study results for impact evaluation of other resource development (notably mining) and for formulating long-term regional development plans for the area."[52] For the Tahltan, qualitative analysis trumped data management: "Meaningful impact assessment has to evaluate the nature of change."[53] The data were intended to facilitate discussion about future land-use opportunities and alternatives rather than to justify current land-use projections and impacts associated with dam construction.

Furthermore, the Tahltan collected and controlled the data. The AUT hired an outside "specialist" to lead each of the sub-sections.[54] However, in many cases it would be Tahltan individuals heading the research teams.[55] Each section was described in terms of study objectives, information needs, methodology, work activities, personnel requirements, and scheduling. The sections were not exclusive categories; information and analytical overlap was built into the research program. Yet even in the assessment of capabilities, options, and effects, the motivation behind the Stikine Basin Resource Analysis was the identification of alternatives and potentials rather than the creation of data for managerial purposes. Furthermore, the "corporate ecology" discourse taken on by the Tahltan in their official dealings with BC Hydro has been sustained. The Cassiar Iskut-Stikine Land and Resources Management Plan of 2000 is the most tangible example: it is an advisory document that leads the Tahltan into discussion with governments and resource corporations on land-use questions.[56] Resource development questions are directed through the Tahltan Heritage Resources Environmental Assessment Team (THREAT). THREAT is involved in all environmental assessment processes within Tahltan territory. Institutional developments like these attest to the long-term impact of Tahltan self-organization in response to potential dam construction.

At the community level, the Tahltan actively resisted BC Hydro's data-gathering activities. There were several violent confrontations between community members and BC Hydro personnel and equipment. In an effort to maintain good relations, BC Hydro never pressed charges. Anti–BC Hydro graffiti was common, and bumper stickers and signage proliferated. The most poignant sign was draped across the river along the Stikine Bridge, which would have been flooded had the dams been built: "Dam the Stikine, Dam the Iskut, Dam the people." There were also isolated acts of arson and sabotage. Eight trailers at the main BC Hydro camp at Bob Quinn Lake were set ablaze, $10,000 of aviation fuel was burned, and BC Hydro employees were fired upon.[57] These confrontations were uncommon, but they do reflect the depth of animosity felt by inhabitants.

Indigenous residents of Telegraph Creek and Iskut were uniformly antagonistic towards Bentley Le Baron, the BC Hydro consultant hired to produce a social and economic impact assessment. His report, submitted in August 1982, was, he acknowledged, hampered by a serious "deficiency": "Native groups in the region have declined to cooperate with the consultant, either to provide statistical data for community profiles or to discuss their particular needs and aspirations as well as attitudes and concerns about the project."[58] His attempts to "arrange a study program that would meet Native needs as well as Hydro's needs" were rebuffed by the leadership of both bands and by George Asp and the executive of the Association of United Tahltans.[59] This was a serious limitation, but Le Baron produced a report regardless.

Individual Tahltan made explicit references to the integral place of the river and animals in Tahltan culture. Speaking in Vancouver at the BC Utilities Commission Rate Hearings in the summer of 1982, elder Henry Tashoots of the Iskut band deplored the intrusive nature of the survey work and road building:

> You see, the traditional trapping ways of our people express that the trapping in the winter was harvested more or less with care, as to the populations they are trapping. Now during the summer months the animals in the area have a chance to re-establish its traditional population. So that it was always regulated. Now if you start sticking … helicopter pads or landing areas in the surrounding area there is a population of mink – now, these are very shy animals – now, what is going to happen is that they're going to migrate to another area, an area where they are more or less not in tune with the environment. So what happens during the winter? They don't know where to find food and they dissipate rather quickly.[60]

Tashoots was aware of the harmful consequence of comparatively minor survey work. By extrapolation, the consequences of a series of dams would be greater, by several orders of magnitude. For men like Tashoots, this was an economic issue as well as an ecological one. Trapping and the hunting and guiding industry were integral components of the local economy. Each would have been curtailed drastically with the flooding of important habitats and established traplines. BC Hydro's argument was that significant economic benefits would compensate for the loss of wildlife-derived revenue, but this ignored the entrenched cultural importance of Tahltan interactions with animals.[61]

The Tahltan were not, and never have been, staunchly anti-development but, rather, keen to ensure that their own social, economic, and environmental outlooks were accounted for by those moving into their territory. Tahltan opposition to dams focused on several overlapping issues: planning activities, adverse effects on the salmon fishery, increased pressure on wildlife, local resentment over decisions made by people who knew little about the region, and a mistrust of BC Hydro, especially its pervasive corporate secrecy. The Tahltan had good reason to feel threatened. BC Hydro explicitly admitted as much in its *Stikine-Iskut Interim Report* of 1982. It sought to create a pro- and anti-development binary by focusing on the rapid increase in the non-Native population, the permanence of development, the differences between value systems, and the alienation of wilderness. According to BC Hydro:

> the project would be beneficial for those people who are part of the wage economy and who perceive that hydroelectric development signifies progress. For those people who depend on the subsistence economy or who place high value on the wilderness, the project would require significant adjustments. The increased population and corresponding increases in competition for land and resources (e.g., fish, wildlife and agricultural land) would increase conflict with the present lifestyles. Moreover, the issue of land claims and its relationship to the development remains critical. The issue has implications not only because of its priority for local and regional native people but also because of the political and legal complexities it represents.[62]

The Tahltan saw the dams as a direct threat to a way of life. Curiously, BC Hydro acknowledged the primacy of land claims but refused to see them as more than superficial elements in the economic and environmental assessment process. The progress–tradition binary discussed above is pervasive in BC Hydro's statements. The sacrifice of an underdeveloped northern watershed, embodied by a subsistence economy and embrace of wilderness values, was emblematic of the tough arithmetic that BC Hydro employed to assess the Stikine-Iskut project and future power needs in general.

Concern about the health of the salmon fishery gained strength during the period of BC Hydro's interest in the watershed. The subsistence fishery was of vital cultural and caloric importance to the Tahltan. In the late 1970s, a commercial fishery was established on the Stikine, eventually stabilizing at twenty-five operating licences running out of two processing facilities, the Stikine Salmon Company and the Great Glacier Co-op. Fishers were hugely concerned about the adverse effects of dam construction on salmon habitat and spawning. BC Hydro moved to quell these concerns, hiring consultant P.J. McCart to conduct fisheries analysis. McCart found that the two major salmon species on the river, the Chinook and the Sockeye, "might" be adversely affected but that "potential losses ... can be largely mitigated provided that appropriate measures are instituted."[63] Dams were to be sited upstream of the furthest known ascent of migrating salmon. Fishers and other locals were skeptical: they asserted that BC Hydro studies neglected to consider the potential impacts of water temperature, turbidity, stream velocity and discharge rates, dispersion of homing odours, egg-fry survival, leaching of toxins, reduction of oxygen and hydrogen, channelization, and the drying of sloughs, among other serious side effects.[64] BC Hydro's public relations employees acknowledged these as possibilities, but refused to comment on them, citing the ongoing studies and speculative nature of the work that had been commissioned.[65] This also raises the question of scientific site selection. As Florian Maurer emphasizes in the *Telkwa Foundation Newsletter,* "BC Hydro [and McCart] selected for its studies only such sites on the lower river, where due to heavily silted glacial tributaries, no spawning can possibly occur."[66] Because the federal department of fisheries and Tahltan assessment resources were stretched so thin, the structural shortfalls of the McCart report were left unchallenged.[67]

A second, more comprehensive report typified how knowledge production around the fishery worked to obscure the findings themselves. Beak Consultants of Richmond, BC, reported that the construction of the Stikine and Iskut dams would have significant impacts on fish and immeasurably alter the river's composition. However, the negative report contained an exculpatory caveat. The authority of Beak Consultants' predictions was contingent on how adequately the models they created could describe and reproduce the physical and biological processes within the estuary and their interactions with each other. The scientific models created to analyze the effects on the river had to exemplify the actual physical landscape more specifically than the original, which was always in flux. Thus, the assessment was predicated on knowledge that the organization did not possess. Therefore, Beak could only recommend further study, and the co-dependency of development and assessment was reinscribed in the Stikine.[68]

The fisheries question also highlights the international components of the debate on hydroelectricity. Though only 10 percent of the Stikine passes through Alaska, most of the Stikine catch is taken in Alaskan waters in the Stikine estuary by Alaskan fishers. Because of this, Alaskans, particularly those in the regional communities of Wrangell and Petersburg, demanded input. In addition, a series of treaties and diplomatic arrangements governed the international management of rivers like the Stikine. The International Joint Commission (IJC) was empowered to adjudicate disputes between Canada and the United States. But the dispute over fish may all have been moot. As Alaska governor Jay Hammond (who just happened to own a fishing cabin at Telegraph Creek) asserted regarding the dams in March of 1982, "I don't think it's warranted. I don't think it's cost-effective. I don't think it's viable and I think it probably should not be built, from everything I've heard."[69]

Of course, there were numerous individual supporters of dam construction in northwest British Columbia, although these voices were often behind the scenes or muted by dissent. The most audible proponents were large-scale industrial operations that required huge amounts of relatively cheap power. For example, the Schaft Creek property, believed to contain one billion tonnes of copper-molybdenum ore, was located near the More Creek dam site and on its own could have used all of the power that would have been generated at that site. There was also a smattering of local support, mostly among the ranks of the small-business community. Elected members of the Regional District of Kitimat-Stikine were also generally favourable to dam construction.[70]

Failure and the Study of Unbuilt Environments

BC Hydro's Stikine-Iskut project is the idiomatic expression of the unbuilt environment – where a highly anticipated and highly contested megaproject was planned and engineered but ultimately not built. It is the knowledge created by dam proponents and the conflicts and contestations that the dams engendered that made the unbuilt environment legible in this case. As a result of talking about how they "knew" the Stikine watershed, developers, contractors, and conservationists frequently afforded wildlife and the river in general a type of environmental currency in efforts to circumscribe the ecological effects of the dams, which were designed to promote the development of the Stikine watershed. Vested groups like the Tahltan, government surveyors and bureaucrats, environmentalists, resource extraction companies, contractors, and recreationists conceived of animals and landscapes in overlapping ways, not least because their voices carried unequal power. Interest groups advocated for notions of environmental resources, property ownership, and usufructory benefits that helped to re-write the status of what an "animal" meant in the Stikine. The representational terrain of animals and fish began to shift as they were recast as scientific objects, development impediments, ecological truths, and cultural-economic symbols. Eventually the associated surveys, environmental assessments, and resource and heritage inventories created such a complex discourse around animals and landscape that the project was abandoned.

Or, more likely, this emerging knowledge around animals and the river was part of a complex calculus of imperatives that led to the abandonment of the project. The influences included environmentalist agitation, Tahltan assertions of control over the watershed, a new regulatory structure under the revised 1980 *Utilities Commission Act,* lowered power-demand projections, unstable economic conditions, and uncertainty about international boundaries and legal responsibilities. The Stikine Canyon and the surrounding plateau remain intact (that is, not under water), but the spectral presence of the five unbuilt dams has changed the way residents and outsiders perceive and use the environment and the many newfound "resources" in the watershed. The process and debate around the dams conceived for the Stikine and Iskut Rivers carried analytical and material weight. Conflicting ideas about progressive and prosperous land use were integral to how place was articulated within environmental and corporate discourse. BC Hydro's production of scientific and technological data made it difficult to conceive of or talk about the river in the same way as before. By quantifying and valuing the river, its contents, and its characteristics, BC Hydro ensured that future discourse about the river would be articulated within that framework of understanding, one that prioritized the scientific and the technical.

The Tahltan have faced centuries of conflict with interlopers over the land and resources that formed the material basis of their social, economic, and cultural lives. Partly because of marginality and scale – lack of access, lack of infrastructure, spatial and temporal dislocations between administrative centres and on-site work – the massive hydroelectric projects envisioned for the Stikine have remained unbuilt. Looking at the program of scientific measurement and the process of data production and management focuses critique on how future projects engage with the environment and inhabitants of future northern landscapes. These unbuilt environments are not simply the unrealized dreams of the state and its development apparatus, or the forgotten, failed schemes to modernize an out-of-the-way place like the Stikine watershed. They are also the product of the many interlocutors working to dislodge accepted notions of assessment, enterprise, and expertise. The Stikine and the Iskut Rivers have remained undammed precisely because of the disagreement about what the river means and how it can best be understood.

4
"Industry for the future": Dome Petroleum and the Afterlives of "Aggressive" Development

LIQUEFIED NATURAL GAS is a fickle but persistent commodity. In the early 1980s, Dome Petroleum, one of the largest corporations in Canada, pursued an ambitious natural gas export scheme that would deliver Western Canadian gas across the Pacific Ocean to ready markets in Japan. Dome claimed to be an industry leader in a relatively new gas-to-liquid technology that allowed natural gas to be fundamentally remade as a new socionatural object. Liquefied natural gas (LNG) created new geographies of exchange and value. It was extracted, piped to processing facilities, frozen, shrunk, and transported to formerly inaccessible distant markets. Company engineers, consultants, and technicians wanted to make natural gas, harnessed from developing gas plays in British Columbia and Alberta and from Dome's pioneering gas-extraction scheme in the Beaufort Sea, mobile by transforming its material form. But moving gas required another, more discursive, shift in orientation. Dome Petroleum personnel also worked towards reframing the meanings of gas within altered political, legal, and administrative frameworks. New narratives of gas combined with new geographies of gas to generate LNG development dreams centred on Dome's preferred export location, Grassy Point, thirty kilometres north of Prince Rupert and adjacent to the community of Lax Kw'alaams, a predominantly Tsimshian (now Coastal Tsimshian Nation) community formerly known as Port Simpson. However, Dome's LNG scheme suffered repeated financial, infrastructural, and regulatory blows and was abandoned before the end of the decade. Like Cassiar, Grassy Point became a footnote in northwest British Columbia's resource history.

Fast-forward thirty years: in September 2011, BC premier Christy Clark pledged an "aggressive approach" to liquefied natural gas (LNG) plant and pipeline development on the province's central coast. Cloaked by the fractious pipeline politics surrounding Enbridge's proposed Northern Gateway Pipeline to Kitimat, the Clark government promoted LNG pipelines and terminals in Kitimat and Prince Rupert. In February 2013, in the run-up to a successful reelection campaign, Clark announced a plan to auction land intended for a reinvigorated LNG cluster at Grassy Point. Four concrete proposals emerged from the province's request for submissions of interest in the Grassy Point area, including bids from SK E and S (Korea), Woodside Petroleum (Australia), a partnership between Imperial Oil and ExxonMobil, alongside a pitch from Nexen. Prospective LNG exports to China, Japan, and South Korea would be at the centre of the government's claims (following a report from Grant Thornton LLP, a leading Canadian accountancy firm) that the rapid escalation of the gas industry could transform the provincial economy, bringing in an eye-popping $1 trillion in revenue and creating over 100,000 jobs.[1] Minister of Natural Gas Development Rich Coleman suggested that "building an LNG export industry is an unprecedented opportunity to create thousands of jobs while supplying Asian markets with the world's cleanest burning fossil fuel. Natural gas is redefining the economic prospects of British Columbians, and our government will ensure these benefits are enjoyed for generations to come."[2] Building, creating, and redefining have become the axioms of the new Western hydrocarbon frontier.

At the end of 2013, the Grassy Point site was green-lit for the second time. Nexen (a recent acquisition of the China National Offshore Oil Corporation), in partnership with INPEX Corporation and JGC Corporation (both of Japan), signed a sole-proponent agreement with the province. Aurora LNG, as the partnership is known, paid $12 million dollars for exclusive rights to develop a 615-hectare site at virtually the same location favoured by Dome. This was an exploratory investment (the agreement can be renewed after one year for an additional $12 million) at a time when the provincial leadership was vigorously promoting LNG potential. Oil and gas companies were cautiously optimistic, eager to explore market possibilities without over-committing. Certainly, the partners in Aurora must believe that Grassy Point is a potentially good location from which to export shale gas from their properties in the Liard, Horn River, and Cordoba basins in British Columbia's northeast.[3] Premier

Clark was ebullient at the announcement, lauding the supposed green credentials of LNG:

> China is looking at reducing its greenhouse gas emissions by 93 megatonnes ... they have a pretty ambitious goal and cannot get there without natural gas. Japan has a 30 per cent energy deficit to fill as they move away from nuclear power, and they cannot get there without LNG. When you add it all up, we have an industry for the future.[4]

Many of world's most powerful energy multinationals appear ready to join the fray. There have been fifteen separate concrete proposals for LNG export, all of which will necessarily include pipelines, service economies, and other infrastructures. Speculation and investment is centred on Kitimat, Prince Rupert, and Grassy Point, but there have been plants proposed for Squamish, Campbell River, Texada Island, and the copper-mining ghost town of Kitsault. Liquefied natural gas (LNG), we are told, must be embraced as a resource for a new green economy – a safe, clean, odourless alternative that can operate as a transition fuel in a new hydrocarbon economy.[5] LNG also has intimate connections to burgeoning new energy and extraction economies in northern British Columbia. This includes links to material resources on the cusp of production (for example, new copper and coal mines in the northwest) and to the technologies required to make extracted materials useful. These technologies take two related forms. First, resources are networked with the material infrastructures that will accompany and facilitate extraction projects (for example, the Northwest Transmission Line, the web of pipelines that will transport "dilbit" and gas across the province, and megaprojects such as BC Hydro's Site C dam). At the same time, in order to be profitably brought into being, resources require the assessment, regulatory, and legislative mechanisms meant to appraise the viability of each project as well as the "cumulative effects" of the new political economy of energy and minerals (for example, the Northern Gateway Pipeline inquiry, new environmental impact assessment protocols, and recently relaxed environmental legislation designed to expedite resource development). Together, the networks enabled by these resources and technologies operate at the centre of provincial government claims of future economic prosperity and its leadership role in global ecological citizenship. When Premier Clark was pressed at the Aurora LNG announcement about the potentially deleterious effects of increased emissions from a new LNG economy and its associated extractive infrastructures, she claimed that "we are doing the world a favour" because the consumption of BC gas in China, Japan, and elsewhere would replace dirty coal.[6]

Many others are dubious about the LNG project for a host of environmental, social, and economic reasons: increased emissions; the massive electricity requirements for liquefaction likely necessitating the construction of new large dams at the public's expense; the skittish nature of natural gas markets; murky economics and humble future price projections; the problem of global overproduction, market gluts, and rampant speculation; and undemocratic decision-making, which compromises the uneasy relations between gas companies and Indigenous peoples and buttresses the cozy relations between energy companies and the state. A chorus of dissent, joining prominent environmental organizations, elements of the press, and some local First Nations, has raised questions about LNG in general but also about the pace and timing of recent developments, especially given that gas is in such abundant supply around the world.[7] Some insist that the projected economics for the province do not add up, and that the rosy climate change benefits touted by the premier amount to not much more than an elaborate "greenwashing" campaign.[8]

This chapter views the creation of LNG resources as social relations by placing current movements of LNG, its infrastructures, and its service constituents alongside a failed attempt to move LNG through the same location thirty years ago. Dome Petroleum sought to build an LNG facility in the early 1980s at Grassy Point in order to export gas under contract to Japan, but the current narrative about the inevitability and opportunity of LNG economies should be underscored by a legacy of negotiation, contestation, and compromise around environmental impact assessment, infrastructure megaproject development, and local involvement in attempts to promote socio-natural improvements. These are important times for a political ecology of energy and mining in Canada, as our governments and corporate giants promote extractive enterprise to drive our modern economies.[9] This political ecology must incorporate history in an expanded and comparative temporal narrative, to see resource development as an embedded historical process as much as an appeal to future possibilities of new extractive regimes.[10] Dome's Western LNG Project, seen alongside the contemporary analogues envisioned for the central coast, fits this profile. The Western LNG Project laid the groundwork for the current LNG groundswell at Grassy Point. For thirty years, it remained an unbuilt environment, but in laying out the particular characteristics of LNG export, Western LNG rendered an extractive vision that has persevered.

Making LNG

Canada needs the Western LNG project, and Dome Petroleum Limited is working very hard to bring it into being.[11]

J.R. van der Linden, "The Western LNG Project"

Stories like the export of LNG from the Pacific Coast exist at the heart of expressions of future economies, future prosperity, and future social arrangements. In Canada, as new energy economies are brought into being, their development is mostly framed as a technological issue.[12] Yet David Nye has undermined this energy technological determinism in his studies on the United States by pressing for more nuanced cultural and social explanations of energy systems.[13] Martin Pasqualetti has demonstrated from a policy and systems perspective that energy innovations should be seen as predominantly social, with important, but ultimately secondary, enabling technological components.[14] Taken together, these two perspectives highlight the importance of previous energy systems and energy schemes, which laid the groundwork for present ambitions and aggressive development scenarios.

The energy landscapes that are produced in order to move gas across oceans are socio-technical formations. LNG should be seen as energy infrastructure, and questions must be asked about LNG's materiality. Extracting, freezing, shrinking, and transporting gas requires new combinations of what Gavin Bridge has termed "material-technological" objects, new gaseous formations that are predicated on the transition of space. Following Bridge, I argue that LNG also has a material, technical, and social life as well as a history that is not inevitable but, rather, an outcome of powerful economic and administrative processes designed to turn potential energy into a resource, to make it possible.[15] LNG is a pivot that brings together the disparate elements of the geographical history of energy development, the social relations of energy regimes, the technological innovation of energy infrastructures, and the managerial agenda of the modern state.

Liquefied natural gas is nature engineered to transcend space, to break the material bounds set by geography and geology. We change the materiality of gas to allow it to move across formerly insurmountable space – in this case, the Pacific Ocean. A very simple version of the making of LNG looks something like this: natural gas is (or will be) extracted from shale formations through the combined processes of hydraulic fracturing (or "fracking") and horizontal drilling. It is then pumped as a gas through envisioned pipelines to coastal liquefaction plants (known as trains), where it is cooled to a temperature of –162 degrees Celsius, at which point it assumes a liquid form and is massively reduced in volume – by some 600 times. In liquid form, the natural gas is pumped into spherical, insulated tanks and transported to points of consumption, where it is re-gasified to be used in heating or in the generation of electricity. This process creates a new "space economy of gas" that connects BC gas, through competition, negotiation, and the global energy market with the sites of the so-called Shale Gas Revolution in the United States, to a new archipelago of exploration installations in the Arctic Ocean and to the emergence of competing LNG economies around the globe, wherever gas has been previously left "stranded" by uncooperative geography.[16]

Liquefied natural gas is the first truly global gas technology, and it is increasingly framed as a transition fuel to new energy and economic prosperity. It is found in increasing abundance around the world. Even if its market value lies in the abstract and speculative rather than in the potential for direct energetic sustenance, emerging global gas plays have ushered in the "golden age of gas."[17] The making of LNG into a global gas commodity is the product of a complex economic and socio-technical alchemy that dates from the mid-1960s and gained commercial momentum in the early 1970s when Japan, an island nation with meagre energy supplies, began importing LNG from Alaska in the name of energy security and reduced air pollution.[18] LNG should be seen as part of a new energy system that has emerged out of technological advances but is firmly rooted in the historical processes of market creation, legal negotiations, social licence production, and government incentives. Energy systems are not inevitable but, rather, an outcome of social, economic, and administrative practices designed to make potential energy into a resource – in effect, to make energy economies possible.[19] And so it is essential to historicize energy, in particular the specific historical moments of energy generation that often appear in conceptual form, separate from their geological and economic contexts. To historicize energy in this way allows us to see the connection between possible and eventual futures, as well as how planned energy environments project industrial possibilities and reinforce development dreams.

The Western LNG Project: Surplus, Diversification, Mobilization

> *The years 1980 and 1981 were the "years of the market" where the principal project activity was centred around obtaining the sales agreement which was signed in its final form March 20, 1982. The year 1982 was, and is, the "year of the regulatory approvals," governments permitting, 1983 will be the year of the engineer when we can complete our design and proceed with the construction.*[20]
>
> *van der Linden, "The Western LNG Project"*

Dome Petroleum built the Western LNG Project from the ground up. The company partnered with Nissho Iwai of Japan (a general trading company with a Canadian subsidiary, NIC Resources) in 1978 and in 1980 announced an agreement in principle to proceed with the project. The scheme was the first of its kind in Canada, involving the export of 2.9 million tonnes of gas annually (154 billion cubic feet) to five Japanese utilities in order to serve fifty million customers beginning in 1986. The federal government and the provincial Social Credit government were staunch supporters of the project, but the economics were tricky from the outset. The Japanese buyers agreed to finance the infrastructure construction component of the $3.5–4 billion capital project (liquefaction plant, shipping terminal, storage facilities, and so on) with a loan of close to $2 billion, at 9.75 percent interest, repayable over ten years from project revenues. The utilities themselves would be responsible for the construction of the regasification plants, and Dome partnered with Westcoast Transmission of Vancouver to take over construction of the $1 billion pipeline needed to service the facilities at Grassy Point. In the wake of these negotiations, the metrics of the deal appeared sound, and Dome negotiated a twenty-year export agreement with its Japanese partners.[21]

Apart from the material infrastructure and financial arrangements, Dome had to steer the project through a regulatory maze involving federal and provincial bodies in British Columbia and Alberta. To do so, Dome personnel produced a rhetorical shift around gas, invoking flexible notions of abundance and fluidity in order to make the gas movable. Dome Petroleum's Western LNG Project was based on the first sale of what was constructed as "surplus" Canadian natural gas to offshore customers. This was gas proven available but in excess of the needs of Canadian domestic and industrial consumers. Surplus gas was the rhetorical counterpart of "stranded" resources, conjuring new visions of economic prosperity and market growth. Speaking to the Canadian Institute of Energy in Calgary in November 1982, Dome vice president E.L. Forgues claimed that "the Japanese LNG market provides an excellent opportunity for Alberta and British Columbia to both diversify and enlarge their export markets for natural gas without sacrificing good economic terms."[22] Within this agenda, LNG development based on surplus gas could broaden both resource production and market economies. But, to adapt a phrase from energy scholar Anna Zalik, in order in make LNG a real possibility, surplus gas would have to be *accomplished* by collapsing geographical limits, stimulating sociotechnical potential, and producing a permissive management regime.[23]

Dome was not the only energy company poised to mobilize the LNG opportunity. In November of 1981, J.A. Carter Energy's Transpac project went public with a competing bid to build a $2.8 billion plant on Ridley Island near Prince Rupert, with a capacity of 500 million cubic feet a day. Then, in early 1982, Petro-Canada announced its intention to compete for the ostensibly limited supply of gas with its Rim Gas project. Carter had been doing pre-feasibility studies over the previous year for a smaller-scale project, while Petro-Canada had begun engineering studies for its site outside of Kitimat – a $2.3 billion project that would start at 250 million cubic feet daily, and eventually triple its output to 756 million cubic feet daily as prospective gas from Western Canada and developing extraction schemes in the Mackenzie Delta came online.[24] (By contrast, the current proposal for Aurora LNG at Grassy Point calls for inputs of 3.7 billion cubic feet per day.[25]) All three of the 1980s LNG projects, intended to export liquefied natural gas to Japan, had to submit detailed proposals to a reconfigured British Columbia Utilities Commission (BCUC) process that required all energy and extraction projects to be reviewed and approved.

Each of the proposals sought to diversify the prospective LNG export into derivative industries. Petrochemical plants were part of all three proposals. Dome, Carter, and Petro-Canada all felt that the inclusion of a chemical plant would improve the chances of success, given that the government was pressing for economic diversification for a province that was suffering an economic depression. Dome also had a petrochemical component, in its plan to build a stripping plant in the northeastern BC gas fields to separate ethane and other "impurities" from the methane desired in the process of making LNG. The ethane would be further altered to produce ethylene as a marketable commodity. Although Dome's Japanese partners were less enthusiastic about the ambitious scope of the petrochemical initiative, Dome's final application included plans for a plant capable of producing 600 million pounds of ethylene annually.[26]

Shipbuilding was another strategic component of each proposal. Dome's original agreement with its Japanese partners contained a potentially sizeable boon for a lagging Canadian shipbuilding industry. Dome would need four (eventually five) specially engineered LNG carriers, each with a 125,000-cubic-metre carrying capacity, making a total of fifty-eight voyages per year.[27] The company estimated the initial construction bill at $2.8 billion, in addition to up-front infrastructure capital and pipeline costs. Dome trumpeted the Canadian content of potential shipbuilding contracts. The tactic initially bore some fruit, as Dome negotiated a preliminary deal with two Japanese firms to transfer LNG tanker technologies to be used in building the LNG tankers in Canada.[28]

Dome already had an established oil and gas exploration regime in the Beaufort Sea, the Arctic Pilot Project in the area north of Tuktoyuktuk. Export to Japan was not Dome's initial vision for the Beaufort exploration and production program; it intended to export gas to the East Coast, and then distribute by pipeline from there. In addition, Dome had entered into partnership with Petro-Canada, Nova Corp, and Melville Shipping to develop long-range plans to coordinate exploration efforts around Arctic Gas projects. By the end of 1981, with the development of the Japanese gas contract, Dome had integrated the Western LNG Project into its Arctic gas dreams, planning to use Arctic gas in its proposal when it came online.[29]

Part of the motivation for the convergence of Dome's Arctic and Pacific gas programs was the looming uncertainty about British Columbia's gas future. By May 1982, the BC government was steering proponents to meet their gas requirements from Alberta suppliers, and pressing them to accelerate their bids to spur the province's sagging economy and job figures. An inquiry by the Ministry of Energy, Mines and Petroleum Development designed to "look for ways to get the best possible use of British Columbia's surplus natural gas," under the leadership of George Govier, was tasked to "identify ... that surplus in a scientific way and to develop ways in which we can continue to develop the surplus over the years."[30] LNG export economies and the prospect of economic diversification depended on the existence of surplus gas, which begged the question: Exactly how much harvestable gas did British Columbia have, and what capacity of gas export and petrochemical industry could it support? The commission estimated that BC gas producers could harness between fifty and seventy-five billion cubic feet per year, not enough to create a "surplus." This provided Dome with the incentive to reorient its gas supply proposals towards Alberta producers. Carter and Petro-Canada maintained plans to find more gas through ambitious exploration campaigns.

Ultimately, the commission pegged British Columbia's harvestable gas potential at 30 trillion cubic feet, although much of this was unavailable with current extraction methods. The commission also affirmed that the expected LNG industry would motivate the discovery and development of new gas resources, lending weight to Carter and Petro-Canada's plans to tie exploration to export.[31] In fact, George Govier ensured that the creation of surplus was contingent on invigorated drilling programs. The report suggested that British Columbia could achieve a surplus of 3 billion cubic metres (about 106 billion cubic feet) of marketable gas if 300 wells per year were drilled. This was no sure thing, as only 110 wells had been drilled the previous year. The provincial government was expecting to reach a decision by the middle of July, leaving the winner to negotiate its bid through the convoluted regulatory network of the National Energy Board and the provincial bodies.

These findings favoured Petro-Canada's relatively modest proposal for an LNG plant at Bish Cove near Kitimat. British Columbia did not have enough surplus gas to support one export facility, so smaller plants were preferred. The Petro-Canada plant had only one train and would process significantly less gas than Dome's proposed plant. BCUC, the provincial body making the recommendation to the energy minister, had to consider the financial feasibility of each project as well, and this weighed heavily against Dome and its $8 billion debt, which might have compromised its ability to follow through on its project. Dome claimed that its debt was tied to other projects, and it was the only bidder with a firm 100-percent sales commitment from Japanese customers and a loan in place to cover the initial plant-construction costs.

On July 15, Minister Bob McClelland announced the selection of Dome over its competitors. McClelland trumpeted four reasons for the selection: a secure position in the Japanese market due to pre-contract arrangements, the highest financial return to the province in corporate taxes and royalty payments, infrastructure investment and jobs, Dome's attractive concessional financing from Japan in the form of guaranteed loans "at very favourable rates," and what the government referred to as its "state of readiness" and general progress of planning and design.[32] The province would benefit to the tune of $3.5 billion dollars and would establish a foothold in the LNG industry until 2005, while developing infrastructure and a culture of extraction and export well into the future. The motivations here were economic – the veneer of social responsibility that we see today was yet to take hold as a narrative mode. Land at Grassy Point would have to be leased from the BC Development Corporation. While Grassy Point was the preferred site, Dome also proposed an alternative site in the Kitimat area near Bish Cove, where Petro-Canada had already done all the work on the ground and had made arrangements to lease land from the Haisla Nation (at the same site as the current Kitimat LNG project proposal). There was considerable lobbying from Kitimat mayor George Thom, who was dismayed that Petro-Canada's bid had been unsuccessful.[33] Dome kept the Kitimat connection at arm's length, though the company saw the allure of projected lower construction costs. Bish Cove was ultimately rejected, however, as Dome was leery of sending LNG tankers on the 125-kilometre voyage through the cluster of islands in the Douglas Channel, the same ocean voyage at the centre of concerns about possible environmental effects of Enbridge's Northern Gateway Pipeline. Politics was at play here as well. Although the provincial government approved the Grassy Point site preferred by Dome, the new energy minister, Brian Smith, claimed that the province might have "jumped the gun." The province pressured Dome to consider the Bish Cove site and hired consultants to consider both sites. Dome, however, held firm in its preference of Grassy Point.[34]

Dome laid out the potential benefits in very simple economic terms. Vice president E.L. Forgues presented a no-risk template for Western LNG: gas prices would be partially tabbed to world energy prices, Dome possessed a 100-percent take or pay contract as well as project financing, Japan was a stable market with no competing local energy sources, and there was "good growth potential for Canadian gas" associated with export to Asia.[35] The company was clearly well positioned for success.

Making Western LNG Possible

As young Engineers coming into the workforce, we envisioned the toughest part of the project would be to design and construct the facility. Times have changed and by far the greatest challenge of the Western LNG Project is to get approval from the two provinces, the federal government and support from other significant interest groups such as the native community.[36]

van der Linden, "The Western LNG Project"

Dome still needed to prove the project's feasibility and set up the framework for community engagement. Dome had considered twenty-six sites for the liquefaction plant, storage facilities, and shipping terminal, before focusing on a shortlist of seven. The shortlist included Bish Cove, but Dome ultimately decided on Grassy Point because of its open sea access, protected harbour, and favourable weather and construction conditions.[37] It was the "technically preferred site on both engineering and environmental grounds."[38] Environmental protection was a "prime objective." Environmental inventories were developed and the company concluded that "the construction and operation of an LNG terminal at Grassy Point on Port Simpson Bay can be accomplished with some minor but acceptable impacts on the environment. Liquefaction of natural gas is a clean process that can exist in harmony with the valuable fishery resources of Port Simpson Bay."[39] Dome was also keen on Grassy Point because of its supposed marginality, an important concept when we historicize how the export of LNG has been framed in British Columbia. In fact, the rationale employed to secure conditions for intervention around LNG have been remarkably consistent – geographical and/or social marginality and subsequent development, simplicity, cleanliness, safety (or absence of risk), and the spectre of surplus or "stranded resources," the great scourge of extractive economies.

Dome certainly employed these terms of reference as it engaged in consultation and produced a legitimate negotiated agreement with the Lax Kw'alaams community at what was then called Port Simpson, the community nearest the planned installations at Grassy Point. This was not a partnership agreement, or a cost-benefit agreement, but an outcome of protracted negotiations between Dome and the Lax Kw'alaams community that dovetailed with the socioeconomic and environmental impact assessment that Dome was required to undertake to prove the feasibility of the project. At the very least, one innovation to emerge from the project was direct engagement between a resource company and local First Nations. This was one of the first arrangements of its kind in Canada, and it occurred without substantial involvement by provincial or federal governments. The Lax Kw'alaams assessed the project, negotiated the agreement conditions and community benefits, and granted the company specific provisions. Yet the process was further complicated because the LNG project was not actually located on reserve land and, as Dome maintained, the company was not legally obligated to negotiate with or compensate the Lax Kw'alaams people, though it did intend to pursue pipeline, road, and power right-of-ways through the reserve. Moreover, the Lax Kw'alaams did not control potential roads or navigational channels to the Grassy Point site and could not stop the expropriation of such territory. Still, the community leveraged its seemingly weak legal standing into an agreement with substantial benefits.

At the outset, Lax Kw'alaams was reluctant to proceed with negotiations without compiling its own data. The band secured funding from two sources: $265,000 from Dome to compile environmental inventories and hire legal representation, and $180,000 from the Department of Indian and Northern Affairs Canada (INAC) through an initiative called the Resource Development Impacts Program.[40] This capital allowed the band to assess the project on its own merits according to community-derived adjudication protocols. At the end of 1982, the Lax Kw'alaams entered into direct negotiations with Dome. The band negotiated payments and contributions of between $15 million and $60 million. This included direct payments – $250,000 per year during construction and $275,000 per year during operations – as well as a host of environmental protection and socioeconomic mechanisms that are often grouped together as mitigation measures. The band secured employment and training provisions, and $400,000 for impact assessments and project monitoring according to the band's own design.[41] Natural gas for commercial and household use would be provided, and, perhaps most importantly for many community members, several capital projects were to be undertaken by Dome, including a road to connect the village to Prince Rupert. Cooperative fisheries monitoring mechanisms were also envisioned, and any damage to fish or wildlife, or their habitats, was to be compensated by the company.[42]

In return, the company secured three important commitments. The Lax Kw'alaams would recognize specific easements needed for the construction of transportation, transmission, and pipeline infrastructure. The band would also support the project as it negotiated the regulatory process. The timing here was crucial. The agreement was signed on October 14, 1983, just days before Dome was due to present material at the NEB hearings into its construction plans. Finally, Dome secured a commitment on land title and future claims that would, in practice, ensure the legal status of the project in perpetuity. The band would support the project even in the event that there was a future territorial conflict between Dome installations and any new lands that might be acquired through comprehensive claims settlements. This was not a waiver of title but rather an insurance policy for Dome's capital projects. This Indigenous/corporate relationship was rare for the early 1980s, and this agreement stood as a bellwether for industry–community relations in British Columbia for some time.[43] The agreement was an economic catalyst for the Lax Kw'alaams in spite of the eventual failure of Western LNG. Various commissioned studies helped the band set up its own logging operation, and in the summer of 1985 the Lax Kw'alaams Development Corporation raised $500,000 to fund an enterprise that exported raw timber and roe-on-kelp to Japan.[44]

Dome's Western LNG may have inaugurated an era of environmental cooperation and circumscribed consultation with First Nations over resource development projects, but the company's environmental engagement went beyond the relationship with the Lax Kw'alaams. Dome produced a specific version of territory at Grassy Point designed to code the site and its environment as marginal, out-of-the-way, and ready for improvement. Construction at the Grassy Point site would have the fewest and least extensive overall impacts – from Dome's perspective this was a valuable characteristic of its geographical marginality. Dome and its consultants harnessed this notion of marginality to shape a vision of Western LNG that would result in "no significant environmental disruption," while any effects "of minor significance" could be easily mitigated by careful management and scheduling of construction protocols and personnel.[45]

Dome could anticipate that any construction "impact on water quality [would be] minor and limited to the period of site preparation" and that a new "roadway can be constructed with minimum environmental effects."[46] The specific ecological conditions of Grassy Point – amid "some of the more subdued landforms found along the northern BC Coast," which were "relatively flat and able to accommodate terminal sites" – were used to buttress the case for development. Birds in the area were of "high resource value" but they were endowed with "relatively low sensitivity." And, in spite of the findings that some animals were "critically dependent on low-land wintering habitat," there was, as the consultants were at pains to emphasize, "no outstanding wildlife capability in the entire study area."[47] Even when effects were unavoidable, or temporarily disruptive, consultants were confident that any environmental disorder could be mitigated. For instance, when pipelines (considered as a separate installation from main plant structures) crossed agricultural land, they could result in "interference with equipment and stock movement, immediate crop loss ... spread of weeds, soil compaction and drainage alteration." These effects could be "minimized, mitigated or ... compensated," while the pipeline structure itself could eventually have "a positive effect by increasing the available grazing area."[48] Indeed, after construction, Dome's plan was to "restore [the environment] to as natural a condition as possible."[49]

Dome's consultants also invoked marginality in addressing the social dislocations and isolation attached to living in a community as out of the way as Port Simpson. Western LNG promised jobs (though very few jobs would go to locals), secondary economic benefits, stability, and modernity. Originally conceived as a resource road to be controlled by Dome, the roadway became a public highway as part of the negotiations with the community. The most singular effect of Dome's project would be the construction of a road linking Port Simpson to Prince Rupert, which would reduce travel and freight costs, provide income from construction and maintenance, allow locals to commute to Prince Rupert to work, create a secondary service economy, and increase potential resource exploitation opportunities.[50] Without the diversification incentive provided by the vital road link, primary employment in forestry, in the fisheries, and in the cannery would remain seasonal. In the view of the consultants hired to assess socioeconomic impacts, this characterized the precarious social and economic reality station of Port Simpson residents.

Consultants produced marginality and its socioeconomic effects: "When fishing is poor, as it was in 1980 and will be in 1981, the total Reserve and all village enterprises are affected. The long seasons of enforced leisure inevitably lead to drinking, irritatives, and flight. These conditions produce family tensions and a constant flow of work for the Band welfare workers." There was also community anxiety around the loss of "traditional" or "cherished" ways of life that operated at the heart of social structures. Problems of housing, band-run social services and policing were sure to surface.[51] Consultants were able to represent (as in Table 4.1) an enumeration and simplification of "social impacts" to be expected in the aftermath of construction. Port Simpson was portrayed as a place that needed Western LNG in order to be brought into the modern world via road links and economic opportunities.

TABLE 4.1
Summary of assessments of social impact: Port Simpson

Category	Impact	Magnitude	Effect	Manageability	Type
Health	Traffic accidents	Small	Scattered	Difficult	Negative
Education	Training programs	Small	Slight	Simple	Positive
Social services	Social assistance decrease	Small	Scattered	Simple	Positive
	Alcohol use	Large	Universal	With effort	Negative
	Child care	Small	Scattered	With effort	Negative
Amenities	Roads	Small	Universal	Simple	Negative
	Stores	Average	Scattered	Simple	Positive
Recreation	Facilities use	Average	Universal	Simple	Positive
	Fraternization	Large	Scattered	With Effort	Positive/ Negative
Community protection	Police	Average	Slight	With Effort	Negative

Source: Reproduced from*Canadian Resourcecon and Inter-Island Coracle Consultation,* "Social and Economic Impacts of the Western LNG Project" (Calgary: Dome Petroleum, 1982), 15.

Against the bright promise of socioeconomic opportunity Dome found little pause in the past. From Dome's perspective, Grassy Point and Port Simpson were places of little environmental, socioeconomic and historical significance. Dome consultants were confident that "only one Indian midden Archeological site" existed in the region of the proposed installations, and that this site could be "salvaged before construction activities commence."[52] Dome was not alone in conjuring up this ahistorical version of Grassy Point. As an explanation for why Dome would not be required to undertake more comprehensive "heritage resources investigations," a BC provincial government research officer suggested that this would "not be a requirement … given the apparent low significance of the site."[53] An apparent lack of significant history influenced the valuation of the site as a resource space.

Lastly, Dome relied on the particular materialities of the LNG to be exported from Grassy Point. Dome characterized LNG as a commodity that entailed very little risk.[54] The company stressed that an "LNG facility is a very clean operation and requires straightforward pollution control technology to produce environmentally acceptable air and water discharges."[55] The LNG produced at Grassy Point would be "very pure liquid methane," which is "non-toxic to plants, animals and fish." The gas would contain negligible amounts of other (presumably more dangerous) hydrocarbons while "virtually all" sulphur, water, and carbon dioxide would be removed on-site.[56] The "very cold nature" of LNG could theoretically pose a freezing risk, but this risk was mitigated because "LNG immediately vapourizes to a gas as it warms [and] does not spread in the same fashion as oil." Nor was the methane to be exported at Grassy Point "spontaneously combustible"; fire was unlikely without an ignition source and as long as the methane concentration in the air remained below 5 percent.[57] Dome's consultants maintained that, because of the mitigation procedures in place, "small leaks or spills of LNG will not affect biological resources." This assertion relied primarily on the supposed cleanliness of the product. Western LNG would contain some nitrogen oxides, but it was sulphur-free. In a rhetorical move that presaged the contemporary messaging about gas, Dome consultants claimed that "natural gas is one of the cleanest burning fuels available and no deterioration of ambient air quality will be expected."[58] If LNG was clear, pure, and safe, if Grassy Point was place of minor ecological, historical, and socioeconomic significance, and if the population and social infrastructure of Port Simpson was in need of improvement, then there could be little objection to Western LNG as a positive, galvanizing source for change.

The Western LNG Project: Failure

A number of uncertainties, however, have caused concern as to the future market for gas.[59]

van der Linden, "The Western LNG Project"

Very quickly things began to go awry. In order to push the project forward, Dome had to secure the cooperation of gas producers in British Columbia, which was a fairly straightforward task. In Alberta, however, producers refused the required compromise of lower netback profits in exchange for the "opening up" of Asian markets. Dome also required approval from regulatory agencies (and governments) in British Columbia and Alberta, as well as from the NEB, and had to fend off the growing exasperation of its Japanese partners and customers, whose gas distribution and infrastructure programs were contingent on the deliveries of LNG from Grassy Point.

By August of 1982, it had emerged that the Japanese utilities could back out of the financing and receiving deal if all the arrangements were not completed by the end of the year. The Japanese partners could also cancel within thirty days if they were not satisfied with the terms of the deal. Dome attempted to use this knowledge as leverage at the NEB hearings for a long-term (at least twenty-year) licence to export gas. In October, economic criticism of the deal began to surface. Donald Armstrong, an economist from McGill University, suggested that instead of being a $4 billion boon to British Columbia's economy as claimed in Dome's cost-benefit analysis, the project would in fact be a net drain, to the tune of $440 million. The same week, a report by Wilfred Gobert, an oil analyst working for the Calgary investment firm Peters and Company, agreed with Armstrong's assessment.[60] These criticisms continued in early 1983, when Dome received tentative project approval from the NEB. Denis Mote of Levesque Beaubien, a Montreal brokerage firm, highlighted the potentially precarious nature of Dome's debt obligations: "I'm very skeptical about the whole project ... Dome is already in a tight financial position. I think the company should sort out its existing problems before rushing headlong into new ventures."[61] Gas producers, particularly in Alberta, were concerned that the high construction and operating costs would mean that their profit margins would be slim, if not compromised completely. This eventuality would also mean lost taxes and royalty revenue for the Alberta government.

Dome's deal with the NEB was conditional. The NEB imposed significant caveats around regulatory approval and the establishment of secure gas supplies. Within the calendar year, Dome had to forge agreements with both provincial governments and their energy regulation boards, but all of these agents were concerned about the accumulation of financial risk. Dome wanted to treat Western LNG installations as "utility conduits" to be operated on a "cost-recovery basis," leaving gas producers to shoulder any economic burden caused by fluctuating energy prices, shipping costs, or construction overruns.[62] This uncertainty was compounded by concerns about Dome's debt. Dome's new vice president John Beddome attempted to forestall much of this critique with assertions about Dome's ability to service its debt and veiled threats that played on the tension between surplus and stranded resources. The alternative to selling LNG to Japan would be to leave it in the ground: there would be an LNG market serviced by Dome, or there would be no LNG market at all.[63] At the end of February, Dome filed the official application, with the goal of reaching production in 3.5 years, with full production of 2.9 million metric tonnes by 1990. This could also be read as a claim for the need for a diversification of markets. As part of an attempted debt-restructuring, Dome publicly contemplated divesting assets, specifically some of its diverse US energy properties, its share holdings in TransCanada Pipelines, and its 12 percent interest in the Alsands oil sands project.[64] Yet the Western LNG Project was pushed forward relentlessly. Dome suggested it would take construction lessons from Syncrude operations in the Athabasca oil sands by building modular pipeline and plant facilities and transporting them to the site at Grassy Point. However, other production assurances fell by the wayside. Dome's earlier Canadian content pledge to the shipbuilding industry was in jeopardy because of concerns about cost and technological aptitude. In addition, as Dome president William Richards suggested during the NEB hearings, the use of Canadian engineers, technicians, and construction workers was increasingly unlikely: "We strongly support the maximum Canadian content for this project, but we draw the line at running any risk whatsoever of hiring a contractor whose competence in these fields may be questionable."[65]

Any optimism generated by Western LNG's conditional NEB approval was short-lived, and the project began to fall apart quickly. Rumblings of discontent came from the Japanese funding partners.[66] Although approvals from provincial energy regulatory agencies were eventually secured (on a 50/50 service basis, adjusted from a 25/75 BC/AB split), Dome still failed to establish the price it would pay British Columbia and Alberta producers for gas. In addition, the Japan National Oil Corporation was apparently reluctant to guarantee the $2 billion loan needed to build the Grassy Point plant and installations, largely because of concerns about Dome's ability to repay. Dome's unique pricing structure also had to be modified to accommodate sinking global energy prices. The pricing formulation imposed by Dome would merge the world oil price and the US border price paid on contracted deliveries. Dome would then subtract its own costs (gas transmission, liquefaction, and return on capital investment), leaving producers and provincial governments whatever was left after Dome's arithmetic was satisfied. The decline in global energy prices would further imperil these profits. In late April, project manager Jerry van der Linden admitted that Dome would have to accept a price $5.81 per million BTU

(British Thermal Unit) instead of the $6.66 per million BTU previously agreed upon.[67] This further endangered Dome's perilous economic situation, and the company was forced to reduce its plant and pipeline construction estimates by 20 percent in order to make up for the new economics of export.[68]

Abruptly, Dome executives began the process of divesting the company from the project, reducing its participation from 90 percent to about 35 percent. The federal government intervened in an attempt to stabilize the project, dispatching energy minister Jean Chrétien to reassure the Japanese partners that Dome was not in financial trouble and, in fact, had the economic backing of the ministry.[69] By the beginning of 1984, already a year behind schedule after several delays due to approval hearings and the difficulty of planning the project itself, Dome executives were travelling to Japan in an attempt to save the project. With a deal that had already been extended three times and was now due to expire at the end of January, Dome still did not have final approval from the NEB. Dome's partners began making arrangements for alternative LNG service, encouraging Australian producers to bring the North Shelf LNG project online early. Meanwhile, other competing global gas producers in Thailand, the USSR, and the US were campaigning to scuttle the Canadian project. An official from one of Dome's Japanese banks stated, "We have given the [Western] LNG project a zero possibility of proceeding."[70] Dome deflected responsibility, suggesting that the delays could be blamed on the "time-consuming" regulatory process, particularly the repeatedly delayed NEB hearings, and the difficulties of securing deals with producers in two different jurisdictions used to the relatively simple negotiations involved in the export of gas to the western US market.[71]

Dome received a one-year reprieve from Chubu Electric, the project's biggest customer (slated to take 1.6 million tonnes of the 2.9 million total), but Chubu also delayed the construction of its own regasification plant in Japan. Cancellation was still an option – it was only because Dome removed itself from the project's lead role that the extension was granted. By November, Dome's stake in the project had been reduced to a minority interest in favour of a partnership between Suncor, Pan-Alberta Gas, and Westcoast Transmission. Canada LNG, as the new partnership was known, granted Dome a small stake in recognition of the investment the company had made over the first years of project development. The NEB granted an export licence to Canada LNG almost immediately, which assuaged the Japanese customers who had advocated strongly for the inclusion of Petro-Canada in the new consortium.

The end of the Western LNG Project was protracted, and there were several corporate reorganizations before its demise. Dome ultimately failed to broker agreements with Alberta producers (who were supposed to supply 50–75 percent of the product) and provincial regulators. The company also failed to convince other Canadian energy companies to commit equity and infrastructural support to the project. This failure turned on two factors: 1) the corollary economic effects of the decidedly mixed performance of Dome's bold attempts to engineer a megaproject for the extraction of oil and gas from the Beaufort Sea, and 2) the company's massive and ultimately debilitating debt after its failed Arctic endeavours and its huge commercial acquisition campaign, which briefly saw Dome become one of Canada's largest corporations but badly overextended the company's capital resources.[72]

Canada LNG failed to bring the project to fruition and gave way in May 1985 to a partnership headed by Petro-Canada and Mobil Oil. The new group brought political leverage (Petro-Can) and LNG expertise (Mobil) but it claimed the deal could only go through if it were offered significant tax reductions and royalty payments. In spite of the intervention of new federal energy minister Pat Carney, Western LNG could not transcend the long-term effects of over-ambitious planning and the short-term effects of a world-wide glut in the LNG market. William Hopper, the chairman of Petro-Canada who had long resisted involvement in the project, suggested, with the clear benefit of hindsight, that LNG itself was fundamentally risky: "Selling liquid natural gas is as tough a way as I know (to sell natural gas). It's as close to brute strength and ignorance as you can get."[73] Western LNG was finally shelved in January 1986. Dome public relations official David Annesley encapsulated the failure in a rare moment of candour: "Obviously we're less than pleased to see a project of that nature die, but the economics couldn't be worked out, I guess."[74] Dome stumbled through the next few years and was eventually bought by Amoco in 1988.

Analogues and Afterlives

Frederick Buell has characterized energy history as necessarily embracing the alternating social currents of exuberance and catastrophe. He suggests that scholars "must conceptualize energy history in terms of a succession of energy systems – systems that are constituted by socio-cultural, economic, environmental and technical relationships" that conspire in particular historical moments.[75] In this spirit, LNG economies on British Columbia's central coast should be seen as part of a sequence of attempts that necessarily build upon one another to remake the energy geographies of the northern half of the province. Western LNG stands as a historical and analytical analogue for present ambitious and aggressive attempts to process natural gas into its liquid form and transport it across the Pacific to ostensibly energy-hungry Asian markets. More broadly, historical analogues such as the Dome case can operate as counter-examples – recovering alternative possibilities that are left obscured in the current development frenzy and illuminating the process that makes new energy regimes possible – or not.

There is, though, another layer to add to the story of hydrocarbon exports from the west coast. In March 1977, Andrew Thompson, a UBC law professor and prominent natural resources lawyer, was appointed to lead the West Coast Oil Ports inquiry to consider a proposed oil port at Kitimat. The inquiry sat for one year, taking testimony up and down the coast, but was terminated before it could render a decision because the proponent pulled out of the project. Yet Thompson still felt compelled to interject into the energy debate. Although there is a tendency to conflate oil and gas in energy-development discussions in Canada – and this should be avoided – the Thompson inquiry addressed not only a specific Kitimat oil export project but all prospective export facilities, as well as broader concerns about shipping hydrocarbons from the west coast. This inquiry, Thompson insisted, "is not merely about the mitigation of adverse environmental, social, and navigational safety impacts – it is about whether an oil port should be built at all!"[76] Energy exports could not be viewed in a vacuum: "logic requires" Thompson argued, "that we address both the energy issues and the environmental and social issues simultaneously."[77] Indeed, the most fundamental question was whether there was even any need for an oil port on the west coast. Dome's Western LNG Project was certainly divorced from the Thompson inquiry – I have found no mention of the inquiry in any company literature or any of the press coverage of the oscillations of Western LNG, and the inquiry itself predates the official announcement of Western LNG. The Thompson inquiry, perhaps emboldened by the 1977 decision rendered by Justice Thomas Berger in the Mackenzie Valley pipeline inquiry, suggested, in a non-binding document not delivered to the House of Commons, that there were sufficient doubts about the safety and efficacy of oil export from the West Coast that the projects should not proceed without further examination.[78]

Turn now to consider the Thompson inquiry alongside the work of the joint review panel of Enbridge's Northern Gateway Pipeline. Both were asked to adjudicate proposals, separated by thirty-five years, to export oil products from Kitimat. In establishing the parameters for assessment, both the Thompson inquiry and the joint review panel have done effectively the same work, asking very similar questions in many of the same communities. The outcomes were quite different, however, with the joint review panel members reporting to federal Natural Resources Minister Joe Oliver in December 2013 that, in spite of considerable opposition from multiple sources, the Northern Gateway should be built, potentially reversing the unofficial moratorium of Pacific tanker traffic maintained in British Columbia for the past thirty-five years.[79] Northern Gateway was approved by the Conservative government in June 2014, though considerable opposition remains.

The historical dimensions of the LNG export story are an important adjunct to Premier Clark's courtship of LNG export economies. At the time of writing, the future energy geographies of Grassy Point and other BC LNG nodes are unclear. The steadfast enthusiasm shown by politicians, gas producers, and prospective exporters belies the uncertainties that challenge all energy infrastructure and extraction initiatives. These uncertainties are exemplified in the failure of Dome's Western LNG Project. By the time this book is published, the immediate question of LNG will perhaps have been determined, and its exact parameters may have been laid out. But regardless of the outcome, questions will surround the energy geographies of northern British Columbia as new projects are envisioned and calibrated to meet the dreams and demands of energy entrepreneurs. We now turn to another energy encounter designed to use energy to transform and transcend space: The Northwest Transmission Line.

5
Transmission: Contesting Energy, Enterprise, and Extraction in the New Northwest Gold Rush

It seems like history is repeating itself, except now it's a transmission line. If the government builds the power line in tough economic times, and the copper industry doesn't come, and the IPPs are slow to come, then it will be just like the 1970s all over again.

Jim Bourquin, Iskut resident and Stikine river guide

THERE IS A NEW RUSH in the Stikine, but this time the scramble is over copper and coal alongside some of the largest known gold deposits in Canada. Despite repeated failures to bring mines to fruition, mining multinationals have persevered over the past twenty years, buying, selling, and negotiating tenure claims to proven underground resources across northwest British Columbia. Teck Resources and NovaGold know that their Galore Creek property, located south of the Iskut River, has a confirmed 786 million tonnes of good-grade copper-gold-silver porphyry.[1] Imperial Metals has begun the extraction of 300 million tonnes of copper-gold ore at Red Chris, just off the Ealue Lake Access Road built by BC Rail in the mid-1970s.[2] To the southeast, Royal Dutch Shell has only recently pulled out of complex negotiations over the removal of 8.1 trillion cubic feet of coal bed methane gas from a 214,000-hectare tenure claim that overlaps with the headwaters of the Stikine, Skeena, and Nass Rivers.[3] An adjacent claim at Arctos Anthracite, owned until recently by Fortune Minerals of London, Ontario, was working to remove 101.7 tonnes of proven anthracite coal reserves to feed industrial operations in growing Asian economies.[4] The scale of these endeavours is massive. But there are at least six other mines, perhaps more, in the area in various stages of the provincial regulatory assessment process. And that is not even counting the various "small-scale" energy projects – run-of-river hydro schemes and wind turbine generating fields – that are (or have been) in various stages of assessment themselves. The Stikine Plateau and its surrounding region has been dubbed "The Golden Triangle."

Mining companies face serious obstacles in converting prospective mineral plays into operational mines. Proponents of mining development often lament the complex regulatory expectations established by the province because meeting them is time-consuming and expensive and could potentially stifle mining investment and development.[5] In the Stikine, mines have faced increasing opposition on environmental and socioeconomic grounds. Mining companies also face anticipated obstacles once the potential ore is out of the ground: the failed transportation lines of the past mean that it is very expensive to ship the ore to smelters or processing facilities. Most importantly, mines of this scale require massive amounts of power to operate, and developed power is lacking in the Stikine. Moreover, the transmission infrastructure that links northern generation sites to southern points of consumption in the province has never been extended west to the Stikine. The transmission of power has become the key limiting factor for the ambitious mining sector in northwest British Columbia, and for regional development more generally.

This brings our attention to the Northwest Transmission Line (NTL). The fate of the NTL has been blanketed in uncertainty since its proposal in 2004, but the goal of diverting more electricity to the northwest has been prominent for at least the past twenty years. The likelihood of the success of the NTL mirrors the ebbs and flows of the mining industry. Debate around the NTL has highlighted recent political history in British Columbia, which reveals a close relationship between mining and party politics. The mining industry has been particularly supportive of the Liberal Party since Gordon Campbell's election in 2001. Harkening back to the halcyon days of the W.A.C. Bennett years, Campbell's legislative agenda facilitated the growth of the mining sector by reducing regulatory requirements and assessment restrictions as well as by promoting the growth of public infrastructure projects. The NTL is a prime example. I assess the promotion of the NTL as a public good and local necessity, reading the claims of "green" power transmission, jobs, and corporate investment against depictions of the line as a public financing of private gain and the environmental and socioeconomic dislocation that often accompanies the movement of international capital into investment hinterlands.

On the face of it, transmission is incidental to the actual environmental effects of mining and the drastic reformulations of social relations that often accompany the extraction of ore bodies. But in British Columbia's northwest, the new economic prosperity promised by mining is contingent on cheap power, and transmission is the central mechanism required for the mining of copper and coal. Debates about the NTL are debates about the future of the region, both the economic possibilities that can be seen in the present mining boom and the threats to the environment and the local economies and social lives tied to it. But these debates are also rooted in the past, in the failures of transportation corridors, in the shadows of envisioned dams, and in abandoned mining landscapes. My aim is to clarify the connections between transmission and extraction, between mining economies and mining environments, and between the economic prospects of the present and legacy of failure that has informed it. In the Stikine, it is the Northwest Transmission Line that makes these connections possible.

Alongside my attempt to historicize northwest British Columbia's extractive economies lies a concern with the politics of science and technology brought to the Stikine to enumerate, calculate, and ascribe value to its environment in preparation for its use. The material produced during the environmental assessment process shows how these materials function as a marker of authority and meaning as new representations of the environment emerge in the Stikine. Ostensibly, environmental impact assessments have been the focus of sustained critical inquiry over the past thirty years.[6] As we will see, however, the gaps and elisions that remain provide ample opportunity for the manipulation of the process to the tangible benefit of those seeking development.

Reviving the Northwest Transmission Line

Premier Gordon Campbell rekindled northern transmission dreams in a September 2008 speech delivered to the BC Union of Municipalities. He pledged $10 million to kick-start the environmental assessment process, to continue community and First Nations consultation, and to resume the scientific and socioeconomic studies that must now accompany major infrastructure projects. Though project costs were estimated at $404 million dollars, Campbell felt justified in proceeding based on the economies that would benefit from electrification. He claimed the NTL would generate $15 billion investment dollars and spawn 11,000 jobs by providing power

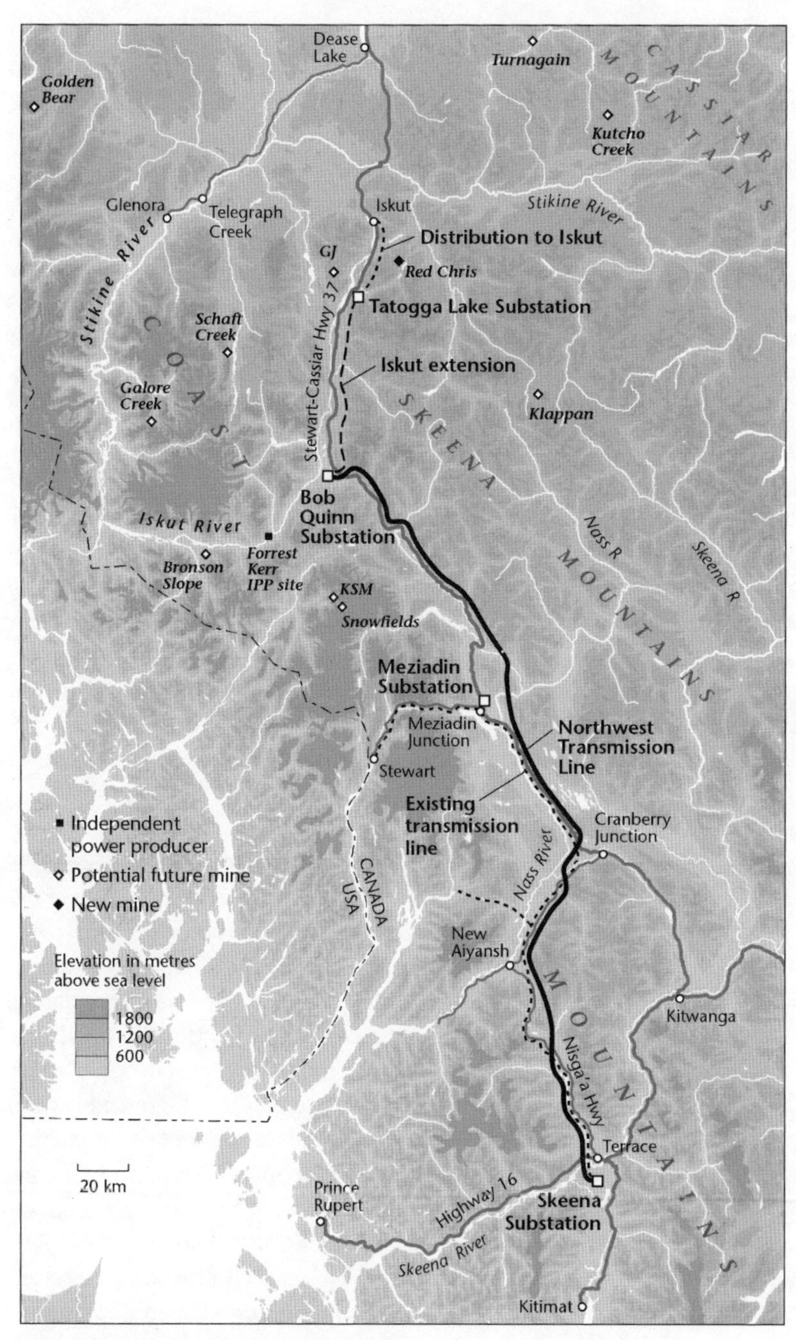

FIGURE 5.1 Northwest Transmission Line Map.
Source: Based on map courtesy of BC Hydro. Adaptation by Eric Leinberger

to ten major mining properties in development in northwest British Columbia.[7] Campbell's projections were taken from a report published on the same day as his announcement by a private mining industry lobby group, the Mining Association of British Columbia.[8] The previous year, a competing industry organization, the Association for Mineral Exploration British Columbia, estimated that regional mineral projects could account for $3.5 billion in investment and two thousand jobs if the projects were built.[9] Campbell was undeterred: the NTL would pay for itself several times over by stimulating prosperity and investment throughout the northwest.

Campbell's government argued that the NTL would be a green energy project. It stressed that, by connecting the region to the grid, the line would reduce greenhouse gas emissions ending northern communities' reliance on dirty diesel generators. This was also a major rhetorical component of an announcement made by Prime Minister Stephen Harper in September 2009. Harper pledged $130 million dollars from Canada's Green Infrastructure Fund towards construction of the NTL. Speaking from Washington, DC, Harper said, "Our government is supporting environmentally-sound infrastructure and initiatives that promote cleaner, greener energy. The Northwest Transmission Line will facilitate the development of green energy and help provide British Columbia's northern and remote communities with more sustainable and affordable power."[10]

Environmental groups expressed skepticism about the green effects of the NTL. The line would run for 344 kilometres from Skeena substation near Terrace to Bob Quinn Lake, essentially following the route of the Nisga'a Highway to Cranberry Junction before joining the Stewart-Cassiar Highway (Hwy 37). By ending at Bob Quinn, the NTL falls short of the communities of Iskut, Dease Lake, and Telegraph Creek, the townsites that were ostensibly to be connected to the grid. Campaigners from various advocacy and policy groups were quick to condemn the plan as "greenwashing." A representative of the Dogwood Institute claimed that "this transmission line is about electrifying coal and metal mines more than it is about clean, green energy." John Horgan, the NDP energy critic, complained that "there's no business plan, there's no private sector partner and there's no environmental approval." In response, Minister for Energy, Mines and Petroleum Resources Blair Lekstrom modulated the green message: "If everyone is unemployed, I don't think the air they breathe is going to matter much." British Columbia Finance Minister Colin Hansen stuck to the financial aspects of the deal: "this is going to open up a quarter of the province." Campbell, Harper, Lekstrom, Hansen, and other proponents of the NTL had allies in the business community. The BC Chamber of Commerce, the Mining Association of BC, mining industry representatives, and various municipal governments all applauded the decision to proceed.[11] Local voices were absent from mainstream accounts but they paralleled media representations of the dispute in their affiliation and temper. Locals may have been muted in the national arena but their polarized views on development, employment, tradition, and the environment reflected the complexity of energy futures in an out-of-the-way place.

How did such fractious debate emerge over a power line and how did Campbell avoid the debate to push the NTL through as a piece of policy and as the focus of a new energy corridor? How did the history of development failure inform planning and discussion over the transmission line? How did the "unbuilt environment," itself a creation that arose out of the failed development projects that characterized the region, become something worth fighting over and fighting for?

Partnerships

The NTL was first imagined in 2004, but the ideas did not gain traction until several years later, when the province (represented by BC Hydro and the BCTC) entered into a funding partnership with NovaGold, the developers of the Galore Creek property. The extent of the Galore Creek deposit had been known for decades, but nobody had managed to develop a cost-effective business plan to deliver the ore to market. Power was the main stumbling block. Smaller mines in the area – Snip, Johnny Mountain, Golden Bear, Eskay Creek – had managed some production success but the scale of Galore Creek, coupled with its location, resulted in projected start-up costs that were too onerous for the junior mining companies that had optioned the property consecutively.

The line was originally part of a plan to connect Galore Creek to the port of Stewart. In October 2007, NovaGold committed $158 million to the construction of the line in exchange for a guaranteed supply of power, needed to build and electrify the mine. Shortly before, in August 2007, NovaGold finalized a 50/50 partnership with mining giant Teck Cominco, a much larger company with the capacity to fund the estimated $2.2 billion mine construction costs. Galore Creek was on track to be the first new operational mine in BC in more than ten years. The availability of cheap power from the proposed line was essential to success. But as Teck began its own assessment process, costs of mine development ballooned. The size of the ore body dictated that the tailings pond would have to be

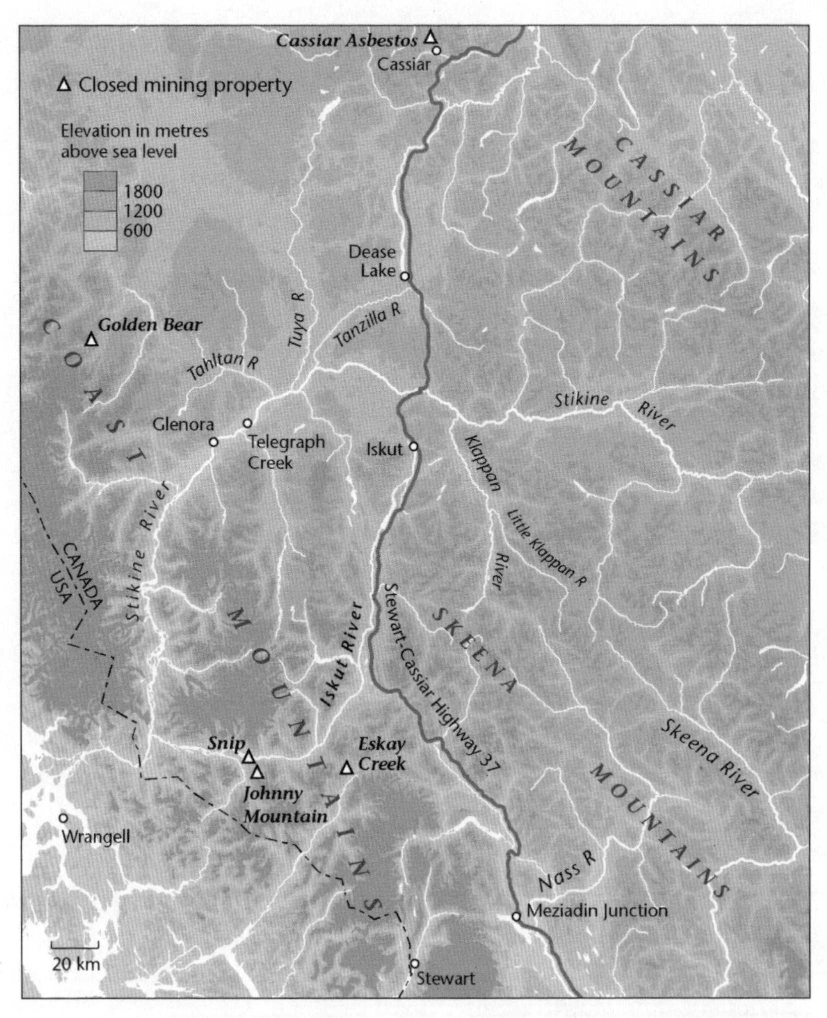

FIGURE 5.2 Closed Mining Properties in the Stikine Area. Map by Eric Leinberger

much larger than previously planned. It would still be located within a recessed glacial valley, but the holding dam would have to be much larger to accommodate the increased tailings volume. There were also concerns about the construction costs for the mine service road that would connect Galore Creek to Highway 37, including controversial plans to build access tunnels of varying lengths. When new construction estimates approached $5 billion, the backers of Galore Creek pulled out of the NTL deal.[12] Officially, the NTL was "on hold," but without a major private backer to offset public financing of the line it was unlikely to proceed.

Mining companies knew that power was essential to the success of the industry in the northwest. The mining industry lobbied for the extension of transmission service into the region. Companies coordinated under the Mining Association of BC (MABC) and the Association of Mining and Exploration of BC (AMEBC), who then partnered with a sequence of northern business groups to pressure the province to forge ahead with construction plans. The Northwest Power Line Coalition (sponsored in part by the MABC) and the Northern Development Initiative Trust led the lobby effort and succeeded in keeping the idea of NTL afloat through the deflated economic years following the termination of the Galore Creek–BC Hydro partnership.[13] In addition, these groups used economic leverage to back the political party most likely to support mining and the large infrastructure spending they were hoping to see built. Lobbyists and mining companies have been very supportive of the BC Liberals, contributing $1.5 million to party coffers and reelection campaigns between 1995 and 2005.[14] Campbell's use of the economic projections of MABC in his assessment of the viability of the NTL suggests the closeness of the relations between industry and government.

By the end of 2009, the NTL, with a fresh injection of capital financing through the Green Infrastructure Fund, seemed poised to go ahead. However, the project still lacked the private–public partnership necessary to build the line. Nevertheless, the province continued the environmental assessment funded by the $10 million grant provided by the Liberal government in September of 2008. The assessment was undertaken by the BC Transmission Corporation (BCTC), an independent Crown corporation that subsequently re-merged with BC Hydro and now functions as an affiliated arm of the utility. An extensive environmental assessment took place over 2009 and early 2010 to supplement the work already undertaken in 2007. The BCTC formally submitted an application for an environmental assessment certificate to the BC Environmental Assessment Office (BCEAO) on April 15, 2010, and it was accepted for review in early June. The BCEAO had 180 days to review the application and make its recommendation to the BC environment and energy ministers, who had to approve before construction could commence. The BCEAO review stopped twice: once in June so that more data could be collected in anticipation of a small route alteration and again in December at the request of federal authorities from Infrastructure Canada, the federal agency supplying the $130 million for construction. The review ended on January 12, 2011, with Minister of Environment Murray Coell and Minister of Mines and Lands Pat Bell expected to issue a certificate, or to reject the proposal, within forty-five days.[15] The process has been fraught from the outset, and should be seen as part of a much broader re-ordering of the practice of environmental assessment in British Columbia and in Canada, designed to facilitate extraction.[16]

In May 2010, while all of the data gathering and consultation was taking place, BC Hydro announced a partnership with AltaGas of Calgary, the new principals for the Forrest Kerr run-of-river hydro project at the confluence of Forrest Kerr Creek and the Iskut River.[17] As part of the $180 million agreement, AltaGas would build its own line to Bob Quinn substation to deliver its power, estimated at a 200-megawatt-generating capacity, to the provincial grid via the NTL. With a new private–public partnership secured and official approval on the horizon, BC Hydro negotiated with First Nations and accepted bids on construction and clearing contracts. After contributions from AltaGas and Infrastructure Canada, the province still had to find the remaining cost (at the time over $250 million due to estimated increases), though the expectations were that much of the total would be covered by agreements with other companies seeking to tap into the power line as mines and independent power producers (IPP) make their own way through the BCEAO process.

Assessments

On the face of it, the BCEAO process is remarkably similar to the self-directed process undertaken by BC Hydro during its attempt to push forward the Stikine-Iskut project thirty years earlier. However, the depth of engagement in environmental and socioeconomic issues and the "legal duty to consult" with interested First Nations clearly sets this process apart from its predecessor. The modern politics of environmental regulation have mechanized the process, while the institutionalization of assessment technologies provides a standardized template for assessment. Rather than a separate document for each component study, assessment is simplified into one document containing everything submitted for consideration by BCTC, which asserts transparency by making the entire document available to the public via the BCEAO website.[18] Transparency is also fostered through deliberate public engagement and consultation. Supplementary catalogues of all of the public commentary from individuals and registered intervenors are available as well. However, the deliberations of BCEAO officials are not part of the public record.

The BCTC application to BCEAO is embedded in an increasingly standardized assessment structure broadly adopted by environmental agencies across the Global North. This format has not escaped the attention of scholars engaged in the critical appraisal of the convergence of science, policy-making, and environmental governance. Issues of scientific authority and expertise are at the forefront of concerns raised by scholars working primarily within political ecology and science and technology studies.[19] Scientists and bureaucrats exercise authority as they activate theoretical and practical knowledge about the natural world and codify information about the threats and risks attached to development. Authority is also challenged by interlocutors (contributors often eager to highlight other ways of knowing) through the foregrounding of scientific uncertainty and its possible links to political and economic interests. Paradoxically, these same interlocutors, by participating in an assessment project not of their own design or intent, reify the authority of science operating in the service of the state. An understanding of science as a social construction, with a destabilized concept of objectivity, has contested the primacy of scientific authority. This has had effects for environmental agencies as they seek to incorporate other ways of knowing into the increasingly uniform models of assessment.[20]

The BCEAO process requires applicants to frame their submissions in a standard format and to collect data using standard methods. This rationalization of scientific behaviour has a profound effect on the way nature is understood as a static historical object in British Columbia. The provincial government tasked BCTC with establishing a baseline view of environmental components. The baseline is the status of present ecosystems and their component parts. Assessment and baseline cataloguing occurs because an alternative environmental use is envisioned. The environmental assessment requires the selection of valued environmental components (VECs): "key biophysical and human features of the environment chosen to focus the assessment on the issues of highest concern and/or relevance to the project."[21] In contrast to the standardized report format, the VEC selection process is unique to each project, incorporating scientific studies, local consultations, and expert and regulator advice as well as input from First Nations and traditional-knowledge studies. Leaving aside the politicized dynamics of these relationships, the VEC selection process as a participatory mechanism is ostensibly inclusive of various forms of knowledge, experience, and the broad spectrum of values attached to environment. The VECs selected for the NTL assessment were fairly comprehensive: atmospheric environment; surface water and groundwater resources; terrain, surficial minerals, and soils; geotechnical stability; fish and aquatic habitat; wetlands; terrestrial ecosystems and vegetation; wildlife and wildlife habitat; archaeology and heritage resources; land and resource use; socioeconomics; visual resources and aesthetics; human health; transportation; and utilities.[22] Most VECs are further reduced into subsections for more detailed analysis. The work is subcontracted to "environmental professionals": engineers, scientists, and bureaucrats with veritable expertise in particular fields relevant to the specific area of study. This conforms to the conventions of environmental assessment, creating a mode of appraisal in which valuation of ecosystems and human–environment relationships is standardized in numbers and normalized in representation.[23] Again, we see the reduction of the complexity of development-driven environmental research.

Each subsection begins with a justification of why the VEC was chosen, demonstrating the relevance of gathered information for environment, social, and cultural lives, regional economies, and public safety. But simplification and generalities obscure complexity. For instance, "atmospheric environment" was chosen "because it encompasses climate and air quality ... [which is] vital to the health of humans, wildlife and vegetation, and influences water quality ... [and] also has aesthetic properties in terms of visibility and odour." The assessment then outlines the mitigating techniques to be employed by contractors: short helicopter trips, full truck loads, speed limit enforcement, equipment maintenance, use of better quality fuels, the application of water to control dust. Potential effects are then summarized. In this case, the reporters concluded there would be no effects beyond the short-term construction phase.[24] This descriptive text exemplifies the power embodied within the definitions of terms of reference. There is nothing "wrong" with the analysis of project effects on air quality, but the boundaries drawn around the analysis preclude a full investigation of those effects. This section of the report does not mention the potential problem of electromagnetic radiation (EMR) to animals, humans, or vegetation, for instance. A full discussion of EMR hazard appears in the section on human health, but it is not highlighted as an environmental hazard. There is considerable disagreement about the long-term health risks of EMR, associated with proximity to long-distance transmission lines.[25] Dust mitigation, clear skies, and greenhouse gases are important, quantifiable, and observable, but radiation – risky, ephemeral, and potentially dangerous – falls outside the environmental purview of the BCTC.

The socioeconomics section illustrates the effects of liberal social values upon the analysis. Socioeconomics "encompasses issues such as employment, education and training opportunities, and community well-being, which were identified through baseline research and consultation with the public, First Nations, and government agencies."[26] The reporters contend that the NTL would generate "benefits" and "economic opportunities" for area residents. In the short term, this is undoubtedly correct, with construction work and contracts confined to the three-year window and an estimated 70 percent of workers coming in from outside the region.[27] There is no discussion of detrimental socioeconomic effects, such as potential social pressures stemming from the influx of temporary workers. Nor is there a discussion of the conceivable socioeconomic effects of mining and other economic endeavours that are likely to follow the line as it opens up the northwest (as in Athabasca or in the area atop the Bakken Shale). Within this framework, development is equated with economic opportunity and the movement towards more fulfilling involvement in an international, market-based economy.

Other sections are more comprehensive. The section on land and resource use "considered (1) how access to land use could be limited or improved by the Project, (2) how the quality of land use activities could be affected, and (3) how forestry activity important to the economy of the region could be supported or limited."[28] Effects during construction would be negligible, "short term and geographically limited." In contrast to assessments for the Dease Lake Extension, there is full acknowledgment that "primary effects" will relate to increased access along right-of-way corridors, though this difficulty, along with "decreased visual quality," were regarded by reporters as "not significant."[29] This section also identified potential effects of "traffic from two mines, growing communities, and increased recreation and tourism," which might have "cumulative effects," though these would be "dependent on timing."[30] Land and resource use changes would be likely, but the particular constellation of effects is unknown, rendering assessment hypothetical.

The section on human health embodies similar characteristics of exclusivity and shortsightedness. EMR (here rendered as electric and magnetic fields [EMFs]) is addressed in this section as an area of public concern. The reporters suggest EMR is found everywhere in daily life, and is therefore inescapable for humans. As EMR becomes naturalized, perceived public risks are sanitized. Reporters predict no adverse health effects on humans but do not mention possible effects on other living bodies.[31] Drinking water and country foods might be affected by sedimentation of streams, metal leaching, acid rock drainage, spilled petroleum products, or "weed control," but a series of procedural safeguards will be implemented to render the "cumulative effects" negligible and localized to construction areas.[32] Again, non-human life forms are not considered within the same framework.

The assessment of risk is historicized even as it takes on a predictive guise in reporters' attempts to understand and mitigate certain environmental effects. "Cumulative effects" is the preferred terminology:

> While an individual effect may be relatively inconsequential, two or more effects that cannot be fully mitigated (termed a residual effect), may combine to produce effects that could be considered significant; this is known as a cumulative effect. As part of the assessment, the anticipated residual effects of the Project were studied to determine if they would interact with the residual effects of the past, present, or probable future projects or activities in the area, and whether this would contribute to measureable cumulative effects.[33]

Past, present, and "reasonably foreseeable" projects come under the purview of BCTC.[34] All of the other proposed or envisioned industrial projects – mines, IPPs, wind power fields, and potential settlement and accompanying human footprints – cannot be accessed within the cumulative effects framework because "their probability of proceeding is not sufficiently known" and there is "insufficient information" to advance possible effect scenarios.[35] The attempt to provide a full review or analysis of cumulative effects is circumscribed by the very economic forces that render these mines economically feasible. The fact that the construction of the NTL has a direct relationship to the potential growth of the mining sector in northwest British Columbia and, therefore, as the progenitor of exacerbated future cumulative effects, is not part of the methodology or analysis. Instead, the study offers a "conceptual best case" scenario, wherein four of the proposed projects (on top of Galore Creek, Red Chris, and the Forrest Kerr IPP) would be up and running by 2015 as long as the NTL was operational on time in 2014. In short, there is no comprehensive mechanism to assess the potential temporal and spatial scales of the effects. Within the cumulative effects assessment (CEA) literature, a distinction is often made between effects (short-term, immediate, and temporary, ideally subject to mitigation) and impacts (longer-term, diffuse, and intractable, subject to uncertainty). In this frame, effects are knowable and therefore actionable, while impacts are expansive and inherently ambivalent across time and space. As Bram Noble has it, "the tyranny of small decisions" that siloes individual projects produces one-dimensional assessments divorced from social and ecological contexts.[36]

Consultation with First Nations and local residents is a central component of the BCEAO assessment apparatus. Relevant First Nations are identified by the BCTC as the seven groups whose rights and title could be affected by the construction of the NTL.[37] The Nisga'a Nation is dealt with separately and is singled out, along with the Tahltan, as the most directly affected Nation. These consultations identified major concerns of First Nations: compliance with traditional laws, physical and biological environmental effects, cumulative effects, project route alternatives, economic benefit and employment, accidents, contamination, emergency response plans, and human health effects. Each concern is addressed in turn in the application, with mitigation and minimization strategies highlighted. The BCTC devoted considerable resources and inaugurated several partnerships to facilitate consultation and education protocols.

Dialogues

Consultations are one way to measure interactions around the NTL. The BCTC considers consultation to be an important procedural element of its assessment apparatus but also a consensus-building mechanism. It has a legal duty to consult and accommodate, based on constitutional precedent and recent case law. In British Columbia, the two central cases involving the Haida and Taku River Tlingit have consolidated the province's duty to consult when there are potential infringements on Aboriginal rights and titles, even if claims, as the Crown maintains, have not yet been proved.[38] *Taku River Tlingit v. British Columbia* involved a dispute over the Tulsequah Chief Mine near Atlin, British Columbia, owned by Redfern Resources. The case confirmed the duty of the Crown (both federal and provincial) to consult but also maintained that various assessment mechanisms had performed that duty appropriately. *Taku River Tlingit* set up the legal framework around consultation and assessment in which the BCTC operates in its dealings with First Nations.

As part of the consultation protocol, the BCEAO documented and posted all of the commentaries it received regarding the NTL terms of reference during the open comments period. These comments have been separated out into an "Aboriginal" comments section and a "public" comments section, underscoring present and historical dislocation in the process. The Tahltan have engaged in the consultation process at the administrative level. Even though it has been confirmed that the touted tangible benefits of grid connection will not be extended to Tahltan communities in the Stikine, the Tahltan have consistently asserted their social, cultural, and economic stake in northwest British Columbia as well as highlighted environmental concerns.

Nalaine Morin, manager at the Tahltan Heritage and Resource Environmental Assessment Team (THREAT) was at the vanguard of this push.[39] Her assessment of the terms of reference (TOR) promoted Tahltan interests and rights in the region. She asserted that "the social, cultural, spiritual and physical well being of the Tahltan people must be respected as guiding ethical principles in consideration of the development and implementation of the environmental assessment [or] ... the Tahltan way of life will be at risk."[40] Morin held the BCEAO accountable for addressing threats to Indigenous rights to land, livelihood, and lifeways:

> The cumulative impacts of the NTL, in combination with other existing or future development activities, on the social, cultural, and economic well-being of the Tahltan are of great concern to us. Our way of life is changing beyond our control in our own territory, in part because the Crown permits development projects without our consent. The BC Environmental Assessment process does not require thorough evaluation of these types of impacts, particularly the social and cultural impacts, and appears to give them lesser weight and/or significance than biophysical environmental impacts. Through our submission, we have highlighted areas of social and cultural impacts for the Tahltan Nation that we believe should be given equal consideration in the environmental assessment process and should be adequately detailed in the Terms of Reference process for the NTL.[41]

The BCEAO did not address these statements directly, instead resorting to the bureaucratic language of the terms of reference themselves:

> BCTC opted into the EA [environmental assessment] process for this proposed project and will consult with all potentially affected First Nations groups as to potential effects of the project. Cumulative effects assessment will be done in accordance with CEA Agency [Canadian Environmental Assessment Agency] guidelines and provincial requirements. Social and cultural effects are addressed in Section 6.12; Cumulative Effects are addressed in Section 10.[42]

In other words, Morin's concerns would be addressed by the BCTC, but the assessment would happen within the framework established by separate and detached government organizations, the BCEAO and the CEA. Moreover, the BCTC emphasized that it chose to participate in the assessment process: it was not obliged to do so, but it submitted to the process in the interest of satisfying public and local apprehension.

Morin raised specific concerns about the obfuscating power of vague language and lack of specificity in reference (or absence of reference) to social and cultural impacts on the Tahltan. She focused attention on important semantic issues: references to Tahltan "communities" should be changed to the Tahltan "Nation" to encompass common identity and territorial cohesion. She asked for clarification on the particularities of structures to be used in the NTL, route selection, and right-of-way appraisal. She showed particular concern about perceived inadequacies in the assessment of potential damage to fish and wildlife populations. Above all, Morin stressed the need for Tahltan involvement and expertise at all levels of the assessment process, in a move that echoed Tahltan claims during the engagement with BC Hydro over the Stikine-Iskut hydroelectric project three decades before.[43]

Morin was also troubled by the cumulative effects protocol to be included in the assessment, basing her judgment on past failures and inadequacies of the process and methodology:

> The Tahltan are well aware of the weakness of the provincial and federal cumulative effects analysis from experience in other EA processes and THREAT could indicate again that unless changes are made to reflect Tahltan concerns, the Tahltan will not recognize the results or mitigations that may result from the analysis. In addition, the section should be based on past, present, and future projects identified by the Tahltan for their territory and the analysis modified to meet social, cultural and environmental criteria of the Tahltan. If an acceptable methodology for cumulative effects can be developed, then the EA Office should ask a third party to conduct the analysis to ensure an open and transparent process.[44]

The BCEAO told Morin, in effect, that she was wrong, substituting the Tahltan experience for a generalized First Nations experience: "First Nations have provided and continue to provide input into the environmental assessment through the technical working group, public comments, traditional use/knowledge, and consultation."[45] Morin succeeded in airing

Tahltan concerns at the outset of the consultation process. She did achieve tangible changes to the terms of reference, and therefore to the assessment process, but was countered by BCTC personnel on some important elements of her critique. Part of the disagreement resulted directly from the institutional nature of the BCEAO. Another element of the disagreement was because of the size of the assessment endeavour and the spread of concerns and interlocutors, while another stemmed from the restricted mandate of the BCEAO. Morin presented the case for Tahltan interests, but it is difficult to get a sense of how individual Tahltan felt about the proposed NTL. A healthy skepticism characterizes relations with the mining industry, though some companies, particularly NovaGold, have made significant inroads in community engagement through consultation, negotiation, and public relations efforts.[46]

In October 2010, BC Hydro's Aboriginal relations and negotiations department published its *First Nations Consultation Summary*. The organization had contracted Golder Associates, a Victoria-based consulting firm specializing in First Nations corporate engagement, to produce the report. The Tahltan section provides an executive-level view of responses to the NTL. BC Hydro was required to undertake this consultation with the Tahltan Nation (represented by the Tahltan Central Council), as well as with other First Nations, as part of the BCEAO process. This duty to consult must be seen within the jurisdictional fragmentation imposed on the Tahltan by the federal and provincial governments through the *Indian Act*. This fragmentation, which requires an artificial divide between long-standing Tahltan leadership and a state-sanctioned governance body in the TCC, produces an ambivalent governance culture. Nevertheless, there is often overlap. Tahltan concerns presented here parallel the concerns raised by Morin in her earlier challenges to the BCTC terms of reference: acknowledgment of socio-cultural values; skepticism about the cumulative effects framework; concern about fish and wildlife, habitats, ecosystems, and wetlands; and general concern about Tahltan involvement in the process. BC Hydro provided funding for the Tahltan to complete a traditional knowledge/traditional use (TK/TU) study, which was included as an appendix in the final draft of the environmental assessment certificate application. BC Hydro seemed receptive to Tahltan concerns regarding cumulative effects. The utility agreed to a supplemental cumulative effects assessment, which was requested by responsible federal authorities – the Department of Fisheries and Oceans, Infrastructure Canada, and the Canadian Environmental Assessment Agency. The report, unavailable to the public, incorporated regional concerns as well as the Tahltan TK/TU study. According to the Golder Associates report, BC Hydro "will take the findings of these reports into account in the development of the Access Plan, construction Environmental Management Plan, and other planning documents."[47]

Public concerns and messages of support for the NTL were also posted by the BCEAO following the publication of the terms of reference. In general, local business people and government representatives wrote in support of the line, stressing the economic opportunity they anticipated could follow in its wake. Lael McKeown of Progressive Ventures suggested that the "project, once completed, will be a transformational project for the future of northwest British Columbia and the province as a whole."[48] Dave Pernarowski, mayor of Terrace and sitting chair of the Northwest Power Line Coalition (NPC), was even more effusive:

> The prospect of this line is one of hope for jobs and sustained economic growth in our area. We believe the NTL would provide much needed high-voltage electricity to the Northwest to spur economic growth and development. Reliable, clean electricity has the power to transform the Northwest economy through mining and other industrial development opportunities, as well as provide employment within the service and support sectors as projects develop and the area's economic base diversifies. We also believe the project will provide an improved standard of living for First Nations and others in the area through improved social and economic benefits.[49]

Overall, the majority of the comments favour the construction of the mine, almost uniformly for the reasons stated above. There was some considerable concern, however, about the environmental cost produced by such a line. These concerns range from a general comment about protecting the earth to more measured commentaries on the potential impacts of the large-scale copper extraction that would be facilitated by the NTL.

While British Columbians debate the merits of the NTL, Alaskans appear ready to embrace the concept and the possibilities it creates for the development of untapped hydroelectric generating capacity in southeast Alaska. The long-mooted southeast Alaska–BC electrical intertie could become a reality if the NTL proceeds. Several Alaskan energy interests (many with ties to Native communities) appear in the public comments minutes: Duff Mitchell (Cascade Creek), Peter M. Naoroz (Kootenoowoo), Robert W. Loescher (Sealaska), and Peter Frisbay (Central Council Tlingit and Haida Tribes of Alaska).[50] Alaskan hydroelectric power interests have come together in an umbrella lobby group of their own, the Alaska–Canada

Energy Coalition (ACE Coalition).[51] There are over eighty potential small-scale hydroelectric sites identified in southeast Alaska, with a total of approximately 3000-megawatt generating capacity.[52] The primary stumbling block for Alaskan energy entrepreneurs is access to the North American grid. The NTL endpoint at Bob Quinn Lake is only eighty kilometres from the nearest potential hydroelectric site in Alaska. The ACE Coalition is particularly strong in Wrangell, at the mouth of the Stikine, where city councillors and the local business community have long sought access to the continental interior through roads and/or an electrical intertie. Of course, Alaskan interests have a staunch ally in the BC mining lobby and its associated public vehicles. Janine North, CEO of the Northern Development Initiative Trust and former co-chair of the Northwest Power Line Coalition (NPC), has offered support for the NTL.[53] She claimed that "northern BC communities were electrified" over the NTL. In North's view:

> Mine development has the strongest potential for economic diversification of central and northwest BC in the short to medium term. Mining and renewable energy projects present an opportunity for sustainable economic development on First Nations traditional lands including joint ventures and contracts that are so important to the supply sector in northern communities, but, only if there is power. Along with other major stimulus investment across northern BC, this is the right investment at a time when it's needed most.[54]

At this point, the social and economic influence of the NTL has reached far beyond northwest British Columbia. The NTL is conceived of as a vital resource conduit in the province and in the international arena. The Tahltan and others in northwest British Columbia face real challenges to ensure that their interests in land, environment, and resources are not subsumed.

Connections

In spite of attempts to address cumulative effects, the assessment failed to consider the consequences of the potential industrial connections that may establish themselves and intersect across the northwest as a result of the availability of cheap power. The environmental outcomes of modern mining projects moving into a region are quantifiable. Earth and rock are displaced, valued minerals are removed, industrial chemicals are processed,

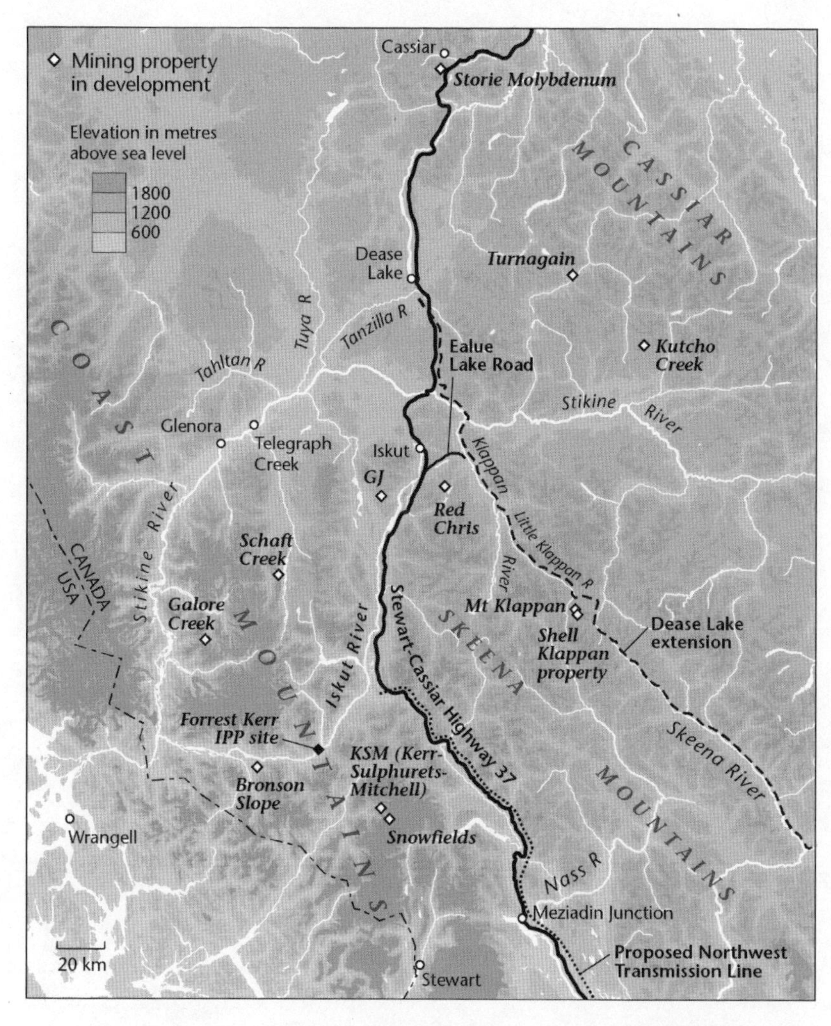

FIGURE 5.3 The Stikine and Area Showing Major Mining Properties, Settlements, the Proposed NTL, Dease Lake Extension, and Highway 37. Map by Eric Leinberger

ecosystems are disrupted and measurable amounts of pollution is produced and accumulates (in the form of acid rock drainage, tailings ponds, and greenhouse gases). The socioeconomic effects of such activities are harder to determine. In the Stikine, there is considerable excitement about the opportunities presented by the new mining economy, just as there is tangible concern regarding the potential social, economic, and environmental dislocations that often accompany the rapid infusion of resource investments – beyond the capacity of local infrastructures to integrate them.

The Tahltan have been building capacity for dealing with the mining industry, particularly over the past ten years. Many Tahltan worked at Cassiar Asbestos, as well as other area mines that had shorter lifespans (Golden Bear, Eskay Creek, Snip, Johnny Mountain). Tahltan marginal employment experience at those mines prompted leadership to evaluate local relationships with mining companies, as well as Tahltan rights and entitlements as they relate to the construction and maintenance of operational mines in Tahltan territory. The first major initiative to address the issues around mining and sustainability was the Tahltan Mining Symposium, held in the summer of 2003, the outcome of which was published the following year under the title *Out of Respect*.[55] The symposium, coordinated by the International Institute for Sustainable Development (IISD), a Winnipeg-based NGO, brought community members together with government officials and mining companies to discuss legacy issues and the reformulation of the previous guiding document, the "Resource Development Policy" (1987). *Out of Respect* set up a template with which the Tahltan could move forward in conjunction with mining development. The stated goals were to ensure that mining progressed in line with Tahltan values, that Tahltan involvement in employment and co-management was assured, and that the Tahltan were guaranteed suitable compensation for work done within Tahltan territory. THREAT was established in 2005 to further facilitate the relationship between the Tahltan and the mining industry.

The Tahltan mining framework has been a vital mechanism for engagement with the mining industry. There are eleven major properties in various stages of environmental assessment. Most are copper mines. Historically, copper prices have fluctuated: in 2013, in the midst of construction, the price of copper averaged over USD$7 per tonne. It has also become potentially profitable to mine the coalfields in the southeast quarter of the Stikine (the Groundhog and Klappan coalfields), first recognized almost a century ago by surveyors from the Geological Survey of Canada.[56] Copper and coal are big business, but a great deal of risk is involved in extraction. Many identified sites have changed hands repeatedly over the years as development dreams fail and are then revitalized (see Table 5.1).

Galore Creek has been at the centre of the push north. Controlled by a jointly owned company, Teck/NovaGold Resources, the project has been revived after its initial postponement in 2007. Economic, environmental, and design feasibility studies are ongoing. NovaGold and the Tahltan Central Council, under then-president Curtis Rattray, negotiated a participation agreement (PA) in 2006, when mine prospects were promising.

TABLE 5.1
Selected Potential Northwest BC Mining Projects

Mine	Company
Galore Creek	NovaGold and Teck Resources
Red Chris	Imperial Metals
Arctos Anthracite Coal	BC Rail (formerly Fortune Minerals and Posco Canada)
Kutcho Creek	Capstone Mining
Storie Molybdenum	Columbia Yukon Explorations
Bronson Slope	SnipGold
Schaft Creek	Copper Fox Metals
Snowfields and Brucejack	Pretium Resources
GJ (Kiniskan)	Skeena Resources
Kerr-Sulphurets-Mitchell (KSM)	Seabridge Gold
Turnagain	Hard Creek Nickel
Foremore	Roca Mines

Note: This is not a comprehensive list. Many potential mining sites have not been included and exploration is ongoing. Listed above are the eleven companies that entered into "earlier-stage" testing (some did later-stage testing as well). These companies operate within the Golden Triangle that has captured the imagination of prospectors and stimulated the interest of investors over the past decade.

The PA contained employment provisions and a guaranteed annual contribution of $1 million dollars to the Tahltan Heritage Trust Fund, to be administered at the discretion of the TCC. The rights of both the Tahltan and NovaGold were set out in the agreement, which recognized Tahltan inherent Aboriginal rights as well as NovaGold's rights to develop their mineral tenure. Though it created some dissent within the Tahltan Nation, the PA was the first concrete application of the *Out of Respect* protocols, though the agreement lapsed when NovaGold's operations were suspended.[57]

The Galore Creek deposit was first discovered in the mid-1950s and was subjected to geological testing and exploration by various companies in the 1960s, 1970s, and 1990s before NovaGold optioned the property in 2003 and developed plans for an open-pit mine to extract copper, gold, and silver. Before the suspension of operations in 2007, the project's design had won approval, though reconfiguration of the mine model required the Galore Creek proponents to re-submit a feasibility study and project execution plan amendments for an updated environmental assessment certificate. The mine life is estimated at eighteen to twenty years, and is accompanied by significant power demands.[58] While debate about the NTL was ongoing, NovaGold expected its pre-feasibility study (fully funded by Teck as per their partnership agreement and completed in mid-2011), to dictate the decision on whether to proceed to permitting and feasibility, depending on the results.[59] Yet still no final decision has been made. Over the long gestation period of the Galore Creek claim, there been some disagreement over road construction, particularly over placement and tunnel construction, but in general NovaGold has enjoyed a productive relationship with the Tahltan over the term of its involvement in the Stikine. In spite of its very considerable potential, Galore Creek remains in a state of suspended animation.

Red Chris has achieved active status but it is also the property that has drawn the most controversy. Imperials Metals acquired Red Chris, located seventeen kilometres from Iskut, from bcMetals in 2007, with a bankable feasibility study already complete and the environmental assessment approved by the BCEAO and federal authorities. This was a "screening" assessment without public input, in which individual components of the project were assessed on their own. The assessment split the project into parts, essentially atomizing both the assessment and the extraction project itself. Impacts from the tailings pond, the campsite, the explosives facility, and the water diversion system were assessed individually, while the mine and mill were not assessed at all. Red Chris is situated on the Todagin Plateau, a critical lambing habitat for the unique Stone Sheep.[60] Todagin lies adjacent to the Spatsizi Plateau Wilderness Park, often referred to as "BC's Serengeti" given its abundant wildlife. It is also the site of traditional Tahltan hunting and fishing grounds.[61]

At the root of the discord over Red Chris is the plan for the disposal and storage of tailings, the accumulated rock sludge/toxic mine waste left over after the ore has been processed. Red Chris plans to build three dams to impound nearby Black Lake (the largest and most central lake in the Ealue Lakes chain) and "reclassify" it into a 2700-hectare tailings dump. The ramifications for Black Lake are obvious, but the likelihood is that damming will also significantly alter local aquatic ecosystems, potentially "killing" adjacent lakes and fish-bearing streams as well. It could also impact the nearby Iskut and Klappan Rivers. The mine is also likely to have significant social impacts. Despite promises to keep workers on site, Imperial Metals has been lodging "hard-to-house" drillers and other temporary personnel at Tatogga Lake, located on the Stewart-Cassiar Highway thirteen kilometres from Iskut. There is considerable concern about socioeconomic impact on the reserve at Iskut.[62] An impact and benefit agreement is being negotiated.

In January 2010, MiningWatch Canada (represented by Ecojustice, a Vancouver-based environmental law firm) filed suit against the federal Department of Fisheries and Oceans, alleging it had breached its assessment responsibilities by not conducting a "comprehensive study" of Red Chris. The Canadian Environmental Assessment Act requires that any metal mine producing over three thousand tonnes of ore per day undergo a comprehensive study, including public consultation. With daily production estimates at thirty thousand tonnes over twenty-five years, Red Chris met the threshold for assessment. Yet because the DFO (and consequently Red Chris) had defined the "project" to exclude the actual mine and focused instead on singular elements of mine infrastructure, no comprehensive study was undertaken. The Canadian Environmental Assessment Agency agreed to the redefinition, and Red Chris was approved on those grounds in 2005. In January 2010, the Supreme Court ruled that the DFO did not act in good faith in its assessment of Red Chris. The loop-hole was closed and future assessment will have to abide more closely to the intention of the *Environmental Assessment Act.* However, MiningWatch's victory was only partial. The Supreme Court's ruling is not retroactive, so Red Chris's environmental assessment certificate, along with the impoundment of Black Lake, remains valid. [63] Red Chris will need forty megawatts from the NTL to process its ore, which it will acquire through a "dedicated line" running 115 kilometres from the mine to the northern terminus of the NTL.[64] In doing so, it will produce an estimated 180 million tonnes of tailings and 300 million tonnes of waste rock over its life span.[65]

Fortune Minerals' Arctos Anthracite is another property that has raised concern: it remains close to being operational after many years of testing and preparation. Elements of this project's transportation strategy have been discussed in the Dease Lake Extension case study. Fortune developed a plan to upgrade the rail bed left after the abandonment of the Dease Lake Rail Extension in order to haul coal to port at Prince Rupert. Fortune released an updated bankable feasibility study in November 2010, suggesting that the Arctos Anthracite deposit at the head of the Little Klappan River contains more anthracite coal than any other source in North America. But coal is cheap; the rail transport plan is just the latest in a long line of use plans developed by the company. Fortune entered into an environmental assessment cooperation agreement with the Tahltan on its way to achieving an environmental assessment certificate.[66] But relations were not always conciliatory. In 2005, Fortune received a court injunction in order to remove a blockade from the access point of Ealue Lake Road. Arrests were made, including the forced removal and temporary incarceration of Iskut elders.[67] This certainly damaged relations between Fortune and the Tahltan, and it remains to be seen how relations will be between between the Tahltan and the mine's new owners, BC Rail.

Shell's coal bed methane (CBM) tenure claim has garnered the most public attention outside the Stikine due to the sensitive ecological dynamics of the area and the potentially serious damage from the controversial gas extraction method to be employed. On February 15, 2011, the provincial government, running against speculation of industry insiders and environmentalists alike, extended the moratorium on coal bed methane exploration in the Klappan. Shell spokesman Larry Lalonde suggested the company had voluntarily decided to continue its suspension of operations in the Sacred Headwaters, which it had first done in December 2008 in the face of mounting international criticism and conflict with the Tahltan. The original 2008 moratorium, intended to last for two years, was designed to build better relations between the parties, to engage in further environmental testing, and to allow the Tahltan to better familiarize themselves with the economic opportunities and environmental hazards inherent in CBM extraction.[68]

Shell began exploration in the Klappan in 2004, drilling three wells in its first year. It was licensed to drill up to fourteen more wells but faced steadfast resistance from Tahltan opposed to CBM development on their territory and from a growing coalition of environmental organizations. The Klabona Keepers Elders Society, a Tahltan group led by Iskut elders, had set up blockades to halt Shell equipment and personnel from entering the Klappan.[69] Originally active in 2005, the society was part of the occupation of the Telegraph Creek Band Office, established in order to remove Chief Jerry Asp. Many Tahltan elders considered Asp's relationship to mining companies, Shell in particular, to be too close.[70] The Klabona Keepers issued its own moratorium to Shell and have asserted their rights through the Klappan Declaration of September 16, 2005.[71] The Skeena Watershed Conservation Coalition (SWCC), along with other metropolitan ENGOs (Dogwood Initiative, Pembina Institute, and Sierra Club BC, among others), called for a much longer moratorium so that the full impacts of CBM extraction could be tested. In the summer of 2011, SWCC partnered with the International League of Conservation Photographers to organize a RAVE (rapid assessment visual expedition), designed to document the region for the publication of a book of photography to be accompanied by an impassioned advocacy text from Wade Davis.[72]

Pembina has led the charge in Canada against CBM extraction and the controversial method of removal, hydraulic fracturing, or fracking. The battle over Shell's plans in the Sacred Headwaters was one of the first environmental encounters to bring the technology of fracking into contemporary public consciousness in Canada. Fracking involves the injection of a mixture of water, sand, and an undisclosed blend of industrial chemicals into wells.[73] The solution cracks the methane-bearing rock, releasing the gas, which is collected at the surface. It is the emblematic "brute force technology."[74] The remaining contaminated water is then pumped back out of the well and dispersed. It often enters back into the water system through groundwater or nearby lakes and streams. There was considerable anxiety that Shell's CBM efforts would contaminate the headwaters of the Stikine, the Skeena and the Nass, jeopardizing three of British Columbia's largest remaining salmon runs. In addition, gas fields have significant surface impacts through the construction of an elaborate system of wells, each of which requires pads about the size of a football field, and the network of roads required to connect the elements of drilling infrastructure together. Shell faced a vigilant, organized, and increasingly well-informed opposition to its methane-based plans in the Klappan. In the final accounting, Tahltan opposition, supported by metropolitan environmental and public policy groups, succeeded in halting Shell's extraction program. The company abandoned its drilling regime in late 2012, although the tenure claim remains potentially open to another company willing to assume the risk. The NTL may not have enough power to bring its extraction goals to light.

These are four of the industrial opportunities on stream in the Stikine. There are several others waiting for power and approval. Schaft Creek rivals Galore Creek in size, likely tapping in to the NTL for 121 megawatts of power to extract its 1.4 billion tonnes of ore. Environmental and economic feasibility studies are underway at Turnagain, where the company originally eyed a 2014 operational start-up that would have required 150 megawatts of power. Kutcho, GJ, KSM, Snowfields, and Bronson are all smaller but have been building mine plans on the availability of power from the NTL.[75] The industrial future of northwest British Columbia is tied directly to the Northwest Transmission Line.

Transmission

> *We can't support this type of project if they are going to come and rape and pillage our territory. With current metal prices, and economic climate changing to positive again, it's a very real concern of ours that 11 projects may come on line and into production by 2016 or 2017. It's a huge concern, and that's why we're working hard at this government to government table to lessen the impacts and decide which projects come on line and which don't.*[76]
>
> *Rick McLean, Chief of the Telegraph Creek band*

None of these mining and energy projects can happen without the Northwest Transmission Line. Mining entrepreneurs have tried and failed to develop power alternatives. The Tahltan negotiated an impact and benefit agreement with BC Hydro, the very nature of which exposes the duality of consequences that are likely to emerge from the NTL. There will be employment, on the line and in the mines that connect to it, and downstream economic effects in Stikine communities in view of the high salaries and new embedded opportunities afforded to Tahltan in the mining sector. Many Tahltan, especially young men, are excited about the prospect of high-paying jobs that would allow them to stay in Tahltan Country.[77] Everyone is also aware of the potential social and cultural dislocations that often trail behind large infusions of investment capital into out-of-the-way places. Anecdotal accounts of abuse and misconduct from Iskut and Tatogga Lake in the summers of 2010 and 2011 were grim reminders of similar events in the days of Golden Bear and Eskay Creek. Many people became temporarily wealthy, but when Barrick Gold shut down Eskay Creek, according to Iskut resident Oscar Dennis, the company took away billions of dollars in mineral wealth and the Tahltan were left empty-handed.[78] There is a constant stream of rhetorical questions in the air at Iskut and Dease Lake: "Where is the ice rink? Where is the community centre? Where is the hospital? Where is the ...?" Residents are asking why those things have never been built and are making sure such community infrastructure and other social benefits are part of the next generation of agreements negotiated with Imperial Metals, Teck/NovaGold, AltaGas, and the others.[79]

The Tahltan have a sophisticated negotiation protocol with firmly defined goals, grounded in an awareness of their Aboriginal rights and title and the geographical scope of their territory. But they face a great deal of uncertainty in terms of how much development will happen and within what timeframe. The Tahltan have some measure of control over which projects move forward. Mining companies need Tahltan approval to proceed, as much from a public relations standpoint as from a legal one. The "stranded resources" of the Stikine, the low-grade copper deposits known about for decades but inaccessible due to high production costs, are suddenly on the cusp of being the epicentre of a new mining boom that could be, alongside LNG, a central component of British Columbia's economy for the next several decades. It would seem that the latest infrastructure dream of the Northwest Transmission Line will ultimately transform the unbuilt environment of northwest British Columbia from one characterized by ambitious failures to one that highlights the escalating profits to be made on commodity markets. The unrealized megaprojects of past years, still visible in the Stikine, still legible even in their immateriality, have allowed the Stikine to remain a virtual place apart. This new megaproject, which stops just short of the watershed boundaries, will bring the Stikine and its residents into much closer connection with global capitalist development and all of the opportunities, risks, and distortions that follow.

For several years, it appeared as though the NTL would remain an unbuilt environment, imagined and engineered but ultimately unrealized as an infrastructure strategy for opening up the Stikine to investment and opportunity. The region would remain unbuilt, until commodity prices and the energy needs of prospective mines conspired to make it necessary in the eyes of policy-makers and the political establishment in BC. The case of the NTL shows that unbuilt environments are not static. They are created by and subject to outside forces at every turn. The knowledge created and the expertise established in one development effort, even if it is ultimately unsuccessful, can be mobilized for following projects.

CONCLUSION

The Tumbling Geography

Two years later, in 1968, I went back again and found the circle of wilderness taking a terrific pasting. The damming and flooding, the logging and road-building, the hundred helicopter bases were leaching it from every angle. Though it was still good ground for a novelist, an alarming number of the oldtimers had been dispersed to hospitals, and my memory of this later summer is a cacophony of get-rich schemes, of white-Indian disparagement and conflict, and Californians immigrating and buying up the homesteads, buying whole chunks of valleys, even to the trading posts and weather stations. There is a frank new air of rapine.

Edward Hoagland, Notes from the Century Before

IN 1966, A YOUNG American writer spent the summer in the Stikine, trying to locate the vestigial remains of a fleeting frontier. In his search for the last wilderness, Edward Hoagland anticipated "the tumbling geography" of development and spent long, bright nights "talking to the doers themselves, the men who no one pays attention to until they are dead, who give mountains their names and who pick the passes that become the freeways."[1] Tucked away in a corner of northern Canada, far from the stratified air of New York City, Hoagland's Stikine was the ultimate moment in the passing of a continental phenomenon, subsumed under development and the progress of the modern. By invoking the frontier mythology, Hoagland deliberately placed himself amid a tradition of American historians and geographers, from Francis Parkman to Frederick Jackson Turner, and the litany of western boosters and tycoons who used the frontier image to lament an idealized and disappearing way of life and to entice newcomers to the newest next, best west.[2] He could hardly have been familiar with the Canadian adaptation of the frontier theories of Harold Innis and J.M.S. Careless, which emphasized the intimate commercial connections between metropolitan centres and the northern resource hinterlands that provided the raw materials that circulated in the emerging market economies.[3] For Hoagland, the Stikine embodied the mythic potential of two North American coordinates, west and north, equally objects of intrigue and investment and, throughout the twentieth century, subject to lasting development pressures and colonial burdens.

The mythic Stikine of the frontier was still tangible in 1966. After the busy years of the Second World War, the Stikine had again lapsed into a sluggish and stoic state that always followed the intermittent years of construction and extraction. The fur trade was still a functioning economy for the Tahltan, the occasional settler still worked the creeks for remnant placer gold, and there were several flourishing hunting and guiding operations. For the most part, however, the watershed was removed from the quicker pace of change that characterized urban cultures and economic relations. Cassiar Asbestos was ramping up production, providing work for adventurous locals and importing labour from afar, but the mine's orientation was elsewhere, facing north through its transportation routes and seeking global markets for its fibres. There were no other operational mines, and the great exploration rush was still a patter. Neither the Dease Lake Extension nor the Stikine-Iskut hydroelectric scheme were under construction. In 1966, Hoagland found a seemingly forgotten landscape, a place apart, just beyond the reach of resource entrepreneurs and the improvement-minded bureaucrats of state institutions. His contemporary Raymond Patterson wrote in the same year that anybody who wanted to see northwest British Columbia as it was had better get there quickly, "before the improvers and the planners – all those who would destroy, recklessly and wastefully, the fair places of the northwest – change the Stikine-Dease watershed out of recognition."[4] When Hoagland returned in 1968, to gather more stories and reacquaint himself with the frontier dreams of eastern mythology, he could feel that the Stikine was on the cusp of profound change.[5] These changes – and the earlier changes that set the conditions of development – have been the subject of this book.

Hoagland recognized that a particular way of life was threatened in the Stikine. But he was mistaken in his rush to declare the closing of the frontier.[6] Like Hoagland, I am most directly concerned with changes to and manipulations of the environment. But in the Stikine, there is a long history of human interactions with nature. Tahltan use of nature long predates the period examined in this book and remains significant. In spite of the Tahltan's historical prominence, the Tahltan positions have not been at the centre of my analysis. In part this absence can be explained by the nature of my sources. I primarily use state documents, corporate data, settler accounts, and environmental consultants' reports to understand the process of environmental change. Where Tahltan accounts are available, I attempt to incorporate them into my analysis; however, these accounts are often only those visible within the same, sanctioned, historical record. The archive is necessarily partial; it is a composite of what has been deemed important to save. Moreover, as in all historical analyses, my own engagements with the archives are subject to my own critical faculties, and my interpretations and recreations of the past. To admit partiality is not to undermine my interpretation and analysis. Instead, it is an acknowledgment of the role of narrative in this type of historical analysis and of my own position as a university researcher engaging with issues of environmental change and ongoing colonialism in an area of British Columbia removed from the heart of provincial economic and political power. I am responsible for the limits and opportunities contained within the knowledge claims made in the project and for the integration of different types of knowledge when the possibility existed. I do not try to speak for the Tahltan; others are in a much better position to do so. Ultimately, my goal has been to interrogate the archives of the Stikine, and to think critically about their many absences and elisions. Archives, in the Stikine as elsewhere, are an arbiter of state power, but they also contain irrepressible voices of resistance that emerge in the cracks and shadows.[7]

The contemporary human and environmental landscape of the Stikine is a reflection of its historical antecedents. Considerable effort has been made to ensure that the Northwest Transmission Line was a successful development initiative; on August 14, 2014, the line was officially opened, albeit at almost double the originally projected cost. Governments, Crown corporations, corporate interests, and First Nations representatives worked together to make sure the necessary work was undertaken to justify the line and to prove its efficacy. Now that the line is a reality, we must wait to see if BC Hydro's "field of dreams" approach to mining development will pay dividends, a situation rendered more complex by a recent global mining downturn. Ultimately, the NTL is a counterpoint to the unbuilt, incomplete, and failed development case studies that push the narrative of this book. I have shown that both "failed" and "successful" megaprojects in the Stikine generate related and linked effects through the planning, negotiation, debate, and investment that go into them. I am interested in the creation of the conditions that make these projects possible or render them unfeasible. My aim has been to show how the tools of developers, improvers, and entrepreneurs have pushed environmental change in the

Stikine. The false dichotomy of failure and success is transcended by the focus on the side effects of development.

The example of the Northwest Transmission Line highlights some curious overlaps with the projects covered in the previous chapters. I have shown how the NTL will facilitate the exponential growth of mining in the area. As Terry Mulligan, president and CEO of the Mining Suppliers Association of BC, points out, "There are at least six mining projects with complete bankable feasibility studies that stand to benefit from access to this transmission line."[8] Mulligan is being modest in his assessment. Many of these six operations could not run profitably without access to cheap power; in any event, it is probable that far more than six projects will benefit from the line. The opportunities for industrial development are likely to have far-reaching consequences for communities and the environment in the Stikine as the exploration, assessment, and enumeration of resources escalates. Elmer Stewart of Copper Fox, the developer of the copper-gold-molybdenum project at Schaft Creek, told a Canadian Press reporter that his company was anticipating the advent of a large, active mining district in northwest British Columbia. Stewart expected that the NTL would usher in a new era of exploration: "Up there a lot of people have focused only on known deposits, which were probably found in the 1960s. I do feel with exploration there are more deposits that will be found."[9]

Improved transportation networks will be a corollary of industrial expansion. There has been significant focus on the rail bed left behind after the abandonment of the Dease Lake Extension. The extension is currently in use as a road by mining and energy companies, which have upgraded sections to facilitate exploration and operations. It has been mooted as a potential resource road to expedite the movement of ore to smelters, refineries, and markets beyond the boundaries of the watershed. Alaskan business interests have suggested that the extension might be a convenient and cheaper pathway for a rail line to connect the northern state to its continental partners. The existence of the raised gravel bed and an established right-of-way, free from the constraints of negotiation, assessment, and compromise, is surely appealing for advocates of resumed construction. There is substantial support in southeast Alaska – particularly in Wrangell – for an electrical intertie through the Bradfield Canal and then through the Iskut valley, just south of the Schaft Creek and Galore Creek mines. This would allow Alaskan power companies to access the North American grid and begin to exploit their considerable potential for small-scale, run-of-river hydroelectric generation. A resource road would likely follow. This hypothetical road has many in the Stikine leery about the loss of revenue that would be experienced if Canadian ports were bypassed by resource traffic, not to mention the potential environmental impacts of a road and transmission line through the Craig River protected area.[10]

Much like the Dease Lake Extension, the Stikine-Iskut project has long been abandoned. However, small-scale hydroelectric schemes are emerging as a significant alternative practice. The Forrest Kerr project will deliver 195 megawatts of "clean energy" to the new BC Hydro substation at Bob Quinn Lake. As construction was completed in 2014, Forrest Kerr built a $180 million transmission line to connect the project to the NTL. AltaGas, the proponent of Forrest Kerr, also plans to build a 55–70 megawatt-generating facility ten kilometres downstream on McLymont Creek. A further 15–18 megawatts is generated from Volcano Creek, also a tributary of the Iskut. BC Hydro may no longer harbour big generating dreams for the area, but these dreams have been replaced by smaller, more logistically manageable and low-profile run-of-river projects that allow water flow to continue unabated and which, proponents argue, have less ruinous environmental consequences. There is an interesting footnote to the story of hydroelectric development in the Stikine. The current Forrest Kerr project is located at the same site as the Forrest Kerr Creek diversion dam originally proposed by BC Hydro in the 1970s. Altagas was able to prove the project's feasibility and secure its environmental assessment certificate partially on the basis of forty years of river-flow data available for the Iskut River. This information was originally compiled as part of BC Hydro's attempts to substantiate the Stikine-Iskut project. In this sense, unbuilt environments can have long-term and unanticipated effects.[11]

Mining and energy projects, and infrastructure developed as a result, will certainly have broad effects on the one economy that has persisted in the Stikine since the days of the Klondike gold rush: the hunting and guiding industry. Family-owned hunting territories are not protected from the effects of development projects. Traplines are similarly threatened. Furthermore, many residents are concerned about the impacts that an exponentially increased and probably itinerant population might have on sensitive habitats and on animal populations. Hunting is an important economic and cultural activity in the Stikine, but it seems clear that the nature of hunting will change fundamentally with the advent of the new resource economy in the region.[12]

Cassiar is still somewhat removed from this new resource frontier as it is situated several hundred kilometres north of the end of the NTL, which terminates at Bob Quinn Lake. However, the little infrastructure that still exists at the former townsite will likely contribute to the further increase in prospecting and exploration predicted for the region. There has been intermittent interest in the contents of the tailings pile, particularly the high concentrations of manganese in the waste rock, while jade mining continues unabated. The underground mine at McDame seems destined to remain closed as health concerns about the use of asbestos are prominent in public discourse and debate. The online presence of the Cassiar community is still strong and continues to broaden through social networking sites and the new photography partnership project with the Northern Archives at UNBC. The Cassiar archives themselves provide enhanced access that will improve research opportunities for interested scholars and the public.

These new developments amount to an unsettling of the unbuilt environment in northwest British Columbia. My project has been to outline environmental and social change in the northwest in the latter half of the twentieth century and to excavate the conditions of possibility that suspended or abandoned development dreams left in their wake. In following the specific development dreams brought to the Stikine, I began to question why these dreams met with failure and what effects those failures had, both on the physical environment and on the way people engaged with nature in the Stikine. My aim has been to show that the conditions of development – all of the assessments and intrigues, the data and debates, the engineering mistakes and community responses – are critical to understanding how and why environmental change occurs and to detailing the breadth and depth of those changes. Failure and success are two sides of the same coin; in terms of development, they are simply different outcomes of the same process. It is worth reiterating, however, that my treatment of the unbuilt environment does not assume that an unbuilt environment is the equivalent of a blank slate. I reject any attempt to use this concept or the analysis provided in these pages to provide cover for, or support, development interests in ways that undermine the interests or rights of First Nations in Canada, the Tahltan in the Stikine or the Lax Kw'alaams on the central coast.

Within the new frontier mythology, development will no longer be unsuccessful in the Stikine. Whereas the Stikine used to be too far removed from the economic and cultural centre to mobilize successfully its resources, it is now perfectly placed on the periphery, ready to be improved enough to play a prominent role in British Columbia's economic future. These case studies have shown that even unbuilt environments have persistent effects. Each significant initiative threatening environmental change has allowed for the growth of ideas about what northwest British Columbia can look like, what it can produce, and what can be taken from it. These studies have highlighted potential connections between northwest British Columbia and other places and between northwest residents and others looking to engage with its environment. Newcomers came to northwest British Columbia to take something out of it, to harness something of value, or to create something that would facilitate movement through its environment. Each time a new development dream was brought north, it added to the material understanding of physical nature of the area. In effect, unbuilt environments reveal as much about the process of development as they do about the material outcomes.

The Stikine remained largely undeveloped in the first decades of the twentieth century, but early development efforts were important because they set up the prospect of improvement and allowed the Stikine to be considered as a place where progress could be achieved. Large-scale industrial development soon followed. Cassiar Asbestos achieved measurable economic success and created a new built environment. The mine's closure and the town's abandonment showed the ephemeral nature of development in the Stikine, challenged the connections that people had forged, and eroded the economic stability of the region. Cassiar was literally un-built when the community was demolished, yet residents found a way to maintain their interest and sense of place by building a new online site of connection that challenged the very materiality of the place.

On the Dease Lake Extension, initial construction achievements were undone by economic and environmental concerns combined with political wrangling and administrative complexities. The extension was abandoned and left unremediated along its route beside the Klappan River. Yet the ecological side effects of its failure have been profound and, perhaps more importantly, the route remains an artery of development in the region. A concurrent development dream, BC Hydro's Stikine-Iskut hydroelectric project, never reached the construction phase. However, more than any of the attempts that preceded it, the Stikine-Iskut project ushered in a process of cataloguing, calculating, and administering the biophysical data of the watershed. The dams were never built, but the planning of them enabled a different way of knowing and assessing the river and its surroundings, one that opposition groups also had to adopt in order to advocate their beliefs about how the river should be used. Variations of these new assessment practices are in evidence in the discourse and debate about the Northwest Transmission Line. For a long time, it appeared that the NTL would remain unbuilt and that, as a result, all of the envisioned mining and industrial developments in northwest British Columbia would again be set aside until a time when local infrastructure could provide adequate support to entrepreneurs and investors. Indeed, it seemed as though the NTL would follow a pattern similar to the other twentieth-century development initiatives in the Stikine. Yet the NTL has gone forward and, when it becomes fully subscribed, it will provide the impetus for the radical alteration of the built environment in the Stikine.

The prospective LNG economy in British Columbia is in a considerable state of flux. The provincial government has continued its resolute cheer-leading of the industry, but doubts have crept into the process, mostly around the financial feasibility of exporting gas from the Pacific Coast of British Columbia when the global gas market is glutted, and while cheaper, more readily available options are coming online elsewhere. By the summer of 2016, there were over twenty distinct proposals for LNG facilities in British Columbia, yet none of these is moving briskly forward with approval, assessment, or construction plans.[13] There are four proposals at Grassy Point, the largest of which – Aurora LNG – would dwarf the capacity of Dome's Western LNG Project by several orders of magnitude. However, all four proponents are non-committal on whether or not they will pursue construction at all, much less provide timelines for completion.

Two recent landmark events have further complicated prospects for industrial development in northern British Columbia. The *Tsilhqot'in Nation* decision in the Supreme Court of Canada, released at the end of June 2014, effectively affirmed the existence of Aboriginal title and imposed fiduciary duties on government and business that go beyond the duty to "consult" and "accommodate."[14] There has been considerable hand wringing, particularly in British Columbia, where few treaties exist, on what this new legal certainty might augur for investment and exploration in the natural resource sector. It remains to be seen how this decision will affect the myriad extraction projects proposed for the northwest, but it does seem certain that First Nations will exercise greater influence on which projects go forward and which are consigned to the unbuilt category. In August 2014, a containment dam at the Mount Polley mine near Quesnel breached, releasing 10 billion litres of water and 4.5 million litres of toxic industrial sludge into nearby waterways. The scale of the accident was massive, enough to make it one of the largest spills of its kind ever recorded. But the disaster had social and economic ramifications further north in the Stikine. Mount Polley is owned by Imperial Metals, the same company poised to begin operations at Red Chris. In fact, the timing could not have been worse for Imperial Metals and Red Chris, which was the property closest to being operational and therefore closest to tapping into the NTL as its main power source (both conditions it has now secured). In the immediate aftermath of the Mount Polley spill, a group of Tahltan elders affiliated with the Klabona Keepers Elders Society set up a blockade of the Red Chris site. Chad Norman Day, in his position as president of the Tahltan Central Council, issued a statement that indicated he has "new questions and concerns" to be addressed to Imperial Metals about the company's plans for the tailings lake at Red Chris.[15] Red Chris does not have all the necessary approvals and only successfully negotiated an impact–benefit agreement with the Tahltan in mid-2015. The framing of the politics of resource development in northwest British Columbia is constantly being adapted, new social meanings and economic imperatives relentlessly broadening the parameters of debate.

More than forty years ago, Edward Hoagland could already discern the frayed edges of the unbuilt environment. Though he lamented the end of a whimsical vision, that potent dream of the frontier, he was perceptive enough to see that the Stikine was on the precipice of social upheaval and environmental challenge. The demise of the frontier took longer than he expected. Yet Hoagland's prescient question remains poignant today: "In the confusion of helicopters and mineral promotions, the question in British Columbia has become the same as everywhere else: How shall we live?"[16] The Stikine is faced with the prospect of change. Its residents are well versed in their own history. They know their environment and have asserted a meaningful voice that can dictate the pace of change. In the Stikine, the question "How shall we live?" remains, but there is opportunity to create a vision of the future that is not simply subject to development dreams. This vision should be informed by the Stikine's difficult past of development failure and be mindful of the possibilities for a prosperous and just future.

APPENDIX

BC Hydro Studies Commissioned for the Stikine-Iskut Hydroelectric Project

Author	*Title*	*Year*
BC Hydro (Hydroelectric Design Division)	Hydroelectric Development of the Stikine River Overview Study	January 1978
Aresco	Preliminary Archaeological Study of the Proposed Stikine-Iskut Hydroelectric Development, Draft Report	April 1980
BC Hydro	Electrical Engineering Transmission from Potential Hydroelectric Generation Projects on the Stikine/Iskut and Liard Rivers	July 1980
BC Hydro	Prospectus: Northern Transmission Studies; Electrical Energy Transmission from Potential Hydroelectric Generation Projects on the Stikine/Iskut and Liard Rivers	July 1980
BC Hydro	Stikine-Iskut Hydroelectric Development: Progress Report on Feasibility Studies	December 1980
BC Hydro	Stikine-Iskut Rivers Hydroelectric Project Environmental Feasibility Studies; Economic Geology	December 1980

Author	*Title*	*Year*
BC Hydro	Stikine-Iskut Hydroelectric Development: Progress Report on Feasibility Studies	December 1980
B.R. Foster and E.Y. Rahs (Mar-Terr Enviro Research)	Relationship Between Mountain Goat Ecology and Proposed Hydroelectric Development on the Stikine River, BC: Final Report on 1979–80 Field Studies	August 1981
Beak Consultants	Preliminary Analysis of the Potential Impact of Hydroelectric Development of the Stikine River System on Biological Resources of the Stikine Estuary for BC Hydro	September 1981
BC Hydro	Stikine-Iskut Hydroelectric Development Feasibility Study: Hydrology, River Regime and Morphology	October 1981
BC Hydro	Stikine-Iskut Hydroelectric Development: Exploration Program and Access Requirements	December 1981
Aspect Consultants and the Association of United Tahltans	Stikine Basin Resource Analysis: An Evaluation of the Impacts of Proposed Hydroelectric Developments, Vol. 1	1981
E.J. Kermode	Climatology of the Iskut-Stikine Basin, BC Hydro, Environmental and Socio-Economic Services	January 1982
Aresco	Stikine-Iskut Hydroelectric Project Heritage Impact Assessment 1980–81	March 1982
P. McCart et al., (Aquatic Environments)	Fish Populations Associated With Proposed Hydroelectric Dams on the Stikine and Iskut Rivers, Vol. 1: Baseline Studies	May 1982
Talisman Land Resource Consultants	Stikine-Iskut Hydroelectric Development Wildlife Habitat Mapping Study	May 1982

Author	*Title*	*Year*
Ian Hayward and Associates	Stikine/Iskut: Preliminary Environmental and Social Impact Assessment of the Stikine-Iskut Transmission System	July 1982
Bentley Le Baron, Canadian Resourcecon	Stikine-Iskut Project: Social and Economic Impacts	August 1982
P. McCart et al., (Aquatic Environments)	Tahltan Fry Enumeration Study	August 1982
BC Hydro	Preliminary Route Engineering Assessment of the Stikine-Iskut Transmission System, Transmission Projects Division	October 1982
Donald Blood and Gary G. Anweiler	Stikine-Iskut Hydroelectric Projects, Supplementary Wildlife Studies	June 1982
Donald Blood, and Gary G. Anweiler	Preliminary Evaluation of Potential Impacts of Stikine Hydroelectric Developments on Wildlife	June 1982
BC Hydro	Stikine-Iskut Development Status Report	October 1982
BC Hydro	Preliminary Planning Report for the Stikine-Iskut Transmission System: A Summary of System Planning, Route Engineering, and Environmental Studies	December 1982
BC Hydro	Stikine-Iskut Hydroelectric Development: Investigations Outline, 1983	1983
Gregory Utzig et al., (Pedalogy Consultants),	Biogeoclamatic Zonation of the Stikine Basin	1983
Wynne Gorman	Habitat Utilization by Ungulates and Other Wildlife in the Proposed Reservoir Areas on the Stikine and Iskut Rivers: A Report of the 1979 and 1980 Field Seasons	1985

Notes

Foreword

1 John Muir's *Travels in Alaska* (Boston: Houghton Mifflin, 1915) includes a version of the essay published as "An Adventure with a Dog and a Glacier," *The Century Magazine* 54, 5 (September 1897), and later expanded into *Stickeen* (1909), available as Project Gutenberg eBook #11673. Julie Cruikshank, *Do Glaciers Listen? Local Knowledge, Colonial Encounters, and Social Imagination* (Vancouver: UBC Press, 2005), particularly Chapter 5.

2 I am indebted to Wade Davis for commenting on this essay and providing some of the details herein. The author of many acclaimed books and scores of scholarly and popular articles, Davis's contributions defy concise summary and his biography has yet to be written. It has been said that his research has been the subject of over a thousand media reports and interviews around the globe, and that it "has inspired numerous documentary films as well as three episodes of the television series, *The X-Files*" (http://www.nationalgeographic.com/explorers/bios/wade-davis/).

See also Leslie Scrivener, *On Ealue Lake: Wade Davis and the Battle for Todagin Mountain,* Star Dispatches Ereads, http://www.stardispatches.com/single.php?id=17.

3 Helen Walters, "Saving a Pristine Backyard Wilderness: Wade Davis at TED2012," TEDBlog, http://blog.ted.com/wade-davis-at-ted2012/; "[Toronto] Star Dispatches: On Ealue Lake: Wade Davis and the Battle for Todagin Mountain," *Toronto Star*, January 3, 2013, http://www.thestar.com/news/insight/2013/01/03/star_dispatches_on_ealue_lake_wade_davis_and_the_battle_for_todagin_mountain.html.

4 Wade Davis et al., *The Sacred Headwaters: The Fight to Save the Stikine, Skeena and Nass*, (Vancouver: Greystone Books, 2011), 18.

5 "[Toronto] Star Dispatches."

6 The "sanctuary" quote is from Tina Loo, *States of Nature: Conserving Canada's Wildlife in the Twentieth Century* (Vancouver: UBC Press, 2006), 194; pages 193 through 201 detail

Walker's efforts to gain protection for the Spatsizi Plateau. For the local account, see Davis et al., *Sacred Headwaters*, 23–24.

7 "A Tycoon Who Wants to Tame a Wilderness," *Life Magazine*, December 9 (1957): 60–64; Lawrence Douglas Taylor, "The Bennett Government's Pacific Northern Railway Project and the Development of British Columbia's 'Hinterland,'" *BC Studies* 175 (Autumn 2012): 35–56.

8 Bev Christensen, *Prince George: Rivers, Railways and Timber* (Burlington, ON: Windsor, 1989). For an account of developments a little further south during this period, see Mary L. McRoberts, "When Good Intentions Fail: A Case of Forest Policy in the British Columbia Interior, 1945–56," *Journal of Forest History* 32, 3 (July 1988): 138–49. For a thought-provoking reflection on Prince George in these years and later, see Brian Fawcett, *Virtual Clearcut: Or, the Way Things Are in My Home Town* (Toronto: Thomas Allen, 2003).

9 Edward Hoagland, *Notes from the Century Before: A Journal from British Columbia* (Toronto: Random House, 1969), 15; George Woodcock, "The Wilderness Observed: A Review Article," *BC Studies* 3 (Fall 1969): 58–63.

10 BCTV, Golden Triangle Eskay Creek Series, 1989, https://youtu.be/_hUitSs_1XU?t=470; Barrick, Eskay Creek Mine, 2004: http://mineralsnorth.ca/pdf/eskay.pdf.

11 Eskay Mining, "The Northwest Transmission Line Project," http://www.eskaymining.com/files/the_northwest_transmission_line_project.pdf.

12 John W. Reps, *Cities of the American West: A History of Frontier Urban Planning* (Princeton, NJ: Princeton University Press, 1979), 276–78.

13 Reps, *Cities of the American West*, 680–82.

14 Timothy F. Flannery, *The Future Eaters: An Ecological History of the Australasian Lands and People* (Chatswood, NSW: Reed, 1994).

15 Wayne Johnston, *The Colony of Unrequited Dreams* (Toronto: Alfred A. Knopf, 1998). Michael Collins, "The Colony of Unrequited Dreams, by Wayne Johnston," on "The page 'Newfoundland Literature' Does Not Exist," https://nfldtxt.com/2014/04/22/the-colony-of-unrequited-dreams-by-wayne-johnston/.

16 BC Spaces for Nature: The Stikine, at: http://www.spacesfornature.org/conserve_stikine.html.

17 Kathryn J. Oberdeck, "Archives of the Unbuilt Environment: Documents and Discourses of Imagined Space in Twentieth-Century Kohler, Wisconsin," in *Archive Stories: Facts, Fictions, and the Writing of History*, ed. Antoinette Burton (Chapel Hill, NC: Duke University Press, 2006), 251–74.

18 Mark A. Greene, review of Burton, *Archive Stories: Facts, Fictions, and the Writing of History*, *Biography* 30, 3 (Summer 2007): 397–400, available at Proquest Literature Online: http://oliterature.proquest.com.fama.us.es/searchFulltext.do?id=R04281283&divLevel=0&area=abell&forward=critref_ft. Some have used the term "visionary architecture" for an architectural design that exists only on paper: see John A. Walker, "Visionary Architecture," Glossary of Art, Architecture and Design since 1945, 3rd ed. (Library Association Publishing, 1992), and Ernest E. Burden, Visionary Architecture: Unbuilt Works of the Imagination (New York, NY: McGraw-Hill,1934).

19 See BC Hydro, Northwest Transmission Line Project, at https://www.bchydro.com/energy-in-bc/projects/ntl.html.

20 Canadian version of the lyrics of a song by Woody Guthrie, as performed by Canadian folk group The Travellers. See http://www.tjff.com/2001/festival_films/the_travellers.html.

21 Hugh Raffles, *In Amazonia: A Natural History* (Princeton, NJ: Princeton University Press, 2002), 7.
22 Collins, "The Colony."
23 Hoagland, *Notes from the Century Before*, 15.
24 Arthur Conan Doyle, "Silver Blaze," *The Memoirs of Sherlock Holmes* (London: George Newnes, 1894), 5–58.
25 Tina Loo, "People in the Way: Modernity, Environment and Society on the Arrow Lakes," *BC Studies* 142, 3 (Summer–Autumn, 2004): 161.
26 Edwin R. Black, "Review of *Pillars of Profit: The Company Province 1934-1972,* by Martin Robin. Toronto, McClelland & Stewart, 1973," *BC Studies* 23 (Autumn 1974): 66.
27 Carl Abbott, *Imagined Frontiers: Contemporary America and Beyond* (Norman: University of Oklahoma Press, 2015), 115.
28 Vicki Albritton and Fredrik Albritton Jonsson, *Green Victorians: The Simple Life in John Ruskin's Lake District* (Chicago: University of Chicago Press, 2016), has a useful, brief discussion of these tendencies. See also Clive Hamilton, *Affluenza: When Too Much Is Never Enough* (Crows Nest, AU: Allen and Unwin, 2005); Bill McKibben, *Earth: Making a Life on a Tough New Planet* (New York, NY: Henry Holt, 2010); Ramachandra Guha, *How Much Should a Person Consume? Environmentalism in India and the United States* (Berkeley: University of California Press, 2006).
29 Wendell Berry was quoted in Will Samson, *Enough: Contentment in an Age of Excess* (Colorado Springs: David C. Cook, 2009), 108; See also Robert Skidelsky and Edward Skidelsky, *How Much Is Enough? Money and the Good Life* (New York, NY: Other Press, 2013).
30 Albritton and Albritton Jonsson, *Green Victorians*, 88.

Introduction: The Stikine Watershed and the Unbuilt Environment

Epigraph: Hugh Brody, "Introduction," in *Stikine: The Great River,* Gary Fiegehan (Vancouver: Douglas and McIntyre, 1991), 13.

1 Letter from Tommy Walker to Winston Mair (Canadian Wildlife Services), British Columbia Archives hereafter BCA), MS-2784, Tommy Walker Fonds, Box 19 File 5, February 17, 1955.
2 Letter from Tommy Walker to Steele Hyland, BCA, MS-2784, Tommy Walker Fonds, Box 17, File 16, January 21, 1960.
3 On Walker, see Tina Loo, *States of Nature: Conserving Canada's Wildlife in the Twentieth Century* (Vancouver: UBC Press, 2006), 192–201.
4 Edward Hoagland, *Notes from the Century Before: A British Columbia Journal* (New York: Ballantine, 1969), 113.
5 David Suzuki, "Foreword," in *The Sacred Headwaters: The Fight to Save the Stikine, Skeena, and Nass,* by Wade Davis (Vancouver: Greystone Books, 2011), viii.
6 John Muir, *Travels in Alaska* (New York: Houghton Mifflin, 1917); Hoagland, *Notes from the Century Before;* Gwen Hayball, *Warburton Pike: An Unassuming Gentleman* (Poole, UK: Gwen Hayball, 1994); Mark Hume, "BC's Sacred Headwaters to Remain Protected from Drilling," *Globe and Mail,* December 18, 2012, http://www.theglobeandmail.com/news/british-columbia/bcs-sacred-headwaters-to-remain-protected-from-drilling/article6504385/.

7 Joel Connelly, "The Last Frontier: B.C. Considers Fate of Its Wild North," *Seattle Post-Intelligencer*, July 8, 1996, A1.
8 John Faustmann, "The Endangered Stikine," *Western Living* 12 (1982): 33–36.
9 Wade Davis, *The Sacred Headwaters: The Fight to Save the Stikine, Skeena, and Nass* (Vancouver: Greystone Books, 2011), 25.
10 Marty Loken et al., *The Stikine River* (Anchorage: Alaska Geographic, 1979), 9.
11 Warburton Pike, *Through the Subarctic Forest* (New York: Arno Press, 1967 [1915]), 62.
12 Bonnie Demerjian, *Roll On! Discovering the Wild Stikine River* (Wrangell, AK: Stikine River Books, 2006).
13 Julie Cruikshank, *Do Glaciers Listen? Local Knowledge, Colonial Encounters, and Social Imagination* (Vancouver: UBC Press, 2010), 7.
14 The term "unbuilt environment" was originally coined by Kathryn Oberdeck to interrogate the planning histories of the company town of Kohler, Wisconsin. Her study highlights the lost opportunities attached to particular visions of community development. Planning history here is necessarily discursive: the polyvalent histories are evident only in their intangible, archival presence. Although these planned spaces never materialized, Oberdeck claims their two-dimensional archival presence can illuminate a collective imagination of what might have been. My aim is to push Oberdeck's central idea further, to test it on the ground, where material effects are as tangible and observable as the imagined effects of failed plans. See K.J. Oberdeck, "Archives of the Unbuilt Environment: Documents and Discourses of Imagined Space in Twentieth-Century Kohler, Wisconsin," in *Archive Stories: Facts, Fictions, and the Writing of History*, ed. A. Burton (Chapel Hill, NC: Duke University Press, 2006), 251–74.
15 Shelley Wright, *Our Ice Is Vanishing/Sikuvut Nunguliqtuq: A History of Inuit, Newcomers, and Climate Change* (Montreal, QC: McGill-Queen's University Press, 2014); Liza Piper, *The Industrial Transformation of Subarctic Canada* (Vancouver: UBC Press, 2009); P. Whitney Lackenbauer and Matt Farish, "The Cold War on Canadian Soil: Militarizing a Northern Environment," *Environmental History* 12, 4 (2007): 921–50; Matt Farish and P. Whitney Lackenbauer, "High Modernism in the North: Planning Frobisher Bay and Inuvik," *Journal of Historical Geography* 35, 3 (2009): 517–44; Andrew Baldwin, Laura Cameron, and Audrey Kobayashi, eds., *Rethinking the Great White North: Race, Nature, and the Historical Geographies of Whiteness in Canada* (Vancouver: UBC Press, 2011); Stephen Bocking, "Science and Spaces in the Northern Environment," *Environmental History* 12, 4 (2007): 868–95; Jody Berland, *North of Empire: Essays on the Cultural Technologies of Space* (Chapel Hill, NC: Duke University Press, 2009); Matt Dyce, "Canada Between the Photograph and the Map: Aerial Photography, Geographical Vision and the State," *Journal of Historical Geography* 39, 1 (January 2013): 69–84; Cruikshank, *Do Glaciers Listen?*
16 John H. Bradbury, "Towards an Alternative Theory of Resource-Based Town Development in Canada," *Economic Geography* 55, 2 (1979): 147–66; Trevor Barnes, "Borderline Communities: Canadian Single Industry Towns, Staples, and Harold Innis," in *B/ordering Space*, ed. Henk Van Houtum, Oliver Kramsch, and Wolfgang Zierhofer (Burlington, UK: Ashgate, 2005), 109–22; Anthony Winson, "The Uneven Development of Canadian Agriculture: Farming in the Maritimes and Ontario," *Canadian Journal of Sociology* 10, 4 (Autumn 1985): 411–38; James Overton, "Uneven Regional Development in Canada: The Case of Newfoundland," *Review of Radical Political Economics* 10, 3 (1978): 106–16.
17 Paul A. David, *Technical Choice, Innovation and Economic Growth: Essays on American and British Experience in the Nineteenth Century* (Cambridge, UK: Cambridge University Press,

1975); Arthur W. Brian, *Increasing Returns and Path Dependence in the Economy* (Ann Arbor: University of Michigan Press, 1994); Paul Krugman, "Increasing Returns and Economic Geography," *Journal of Political Economy* 99, 3 (1991): 483–99.

18 William J. Turkel, *The Archive of Place: Unearthing the Pasts of the Chilcotin Plateau* (Vancouver: UBC Press, 2007).

19 Thomas Berger, *Northern Frontier, Northern Homeland: The Report of the Mackenzie Valley Pipeline,* vol. 1 (Ottawa: Supply and Services Canada, 1977); Hugh Brody, *Maps and Dreams: Indians and the British Columbia Frontier* (Vancouver: Douglas and McIntyre, 1988); Paul Sabin, "Voices from the Hydrocarbon Frontier: Canada's Mackenzie Valley Pipeline Inquiry (1974–1977)," *Environmental History Review* 19, 1 (1995): 17–48.

20 For a contemporary example that is heavy on hubris but largely ignores historical context, see John Van Nostrand, "If We Build It, They Will Stay," *The Walrus* (September 2014), http://thewalrus.ca/if-we-build-it-they-will-stay/.

21 Roxanne Willis, *Alaska's Place in the West: From the Last Frontier to the Last Great Wilderness* (Lawrence: University of Kansas Press, 2010); Paul Brooks, "The Plot to Drown Alaska," *Harper's Magazine* 215 (May 1965): 53–59.

22 Claus M. Naske, "The Taiya Project," *BC Studies* 91/92 (1991): 5–50.

23 Kerry Abel and Ken Coates, *Northern Visions: New Perspectives on the North in Canadian History* (Peterborough, UK: Broadview Press, 2001). A more recent and more deliberately "global" and "environmental" framework in a similar vein is the analytical notion of "northscapes." See Dolly Jorgensen and Sverker Sorlin, eds. *Northscapes: History, Technology, and the Making of Northern Environments* (Vancouver: UBC Press, 2013).

24 For a comprehensive list of megaprojects in the Canadian subarctic and provincial norths in the period between the Second World War and the late 1980s, see Nancy Knight et al., *What We Know about the Socio-Economic Impacts of Canadian Megaprojects: An Annotated Bibliography of Post-Project Studies* (University of British Columbia, Centre for Human Settlements, 1994).

25 Turkel, *Archive of Place*; Claire Campbell, *Shaped by the West Wind: Nature and History in Georgian Bay* (Vancouver: UBC Press, 2005); John Sandlos, *Hunters at the Margins: Native Peoples and Wildlife Conservation in the Northwest Territories* (Vancouver: UBC Press, 2007): Piper, *Industrial Transformation;* Hans Carlson, *Home Is the Hunter: The James Bay Cree and Their Land* (Vancouver: UBC Press, 2008); Shannon Stunden Bower, *Wet Prairie: People, Land, and Water in Agricultural Manitoba* (Vancouver: UBC Press, 2011); Dean Bavington, *Managed Annihilation: An Unnatural History of the Newfoundland Cod Collapse* (Vancouver: UBC Press, 2010); Jamie Linton, *What Is Water? A History of a Modern Abstraction* (Vancouver: UBC Press, 2010).

26 Arn Keeling and John Sandlos, "Environmental Justice Goes Underground? Historical Notes from Canada's Northern Mining Frontier," *Environmental Justice* 2, 3 (2009): 117–25.

27 Raymond Bryant and Sinead Bailey, *Third World Political Ecology: An Introduction* (New York: Routledge, 1997); Raymond Bryant, "Political Ecology: An Emerging Research Agenda in Third-World Studies," *Political Geography* 11, 1 (January 1992): 12–36; Richard Peet and Michael Watts, eds., *Liberation Ecologies: Environment, Development, Social Movements* (London and New York: Routledge, 1996); Piers M. Blaikie and Harold Brookfield, eds., *Land Degradation and Society* (London and New York: Methuen, 1987).

28 Richard Peet and Michael J. Watts, "Introduction: Development Theory and Environment in an Age of Market Triumphalism," *Economic Geography* 68, 3 (1993): 227–53; Rachel Silvey and Katherine Rankin, "Development Geography: Critical Development Studies and

Political Geographic Imaginaries," *Progress in Human Geography* 35, 5 (October 2011): 696–704.

29 Anna Tsing, *Friction: An Ethnography of Global Connection* (Princeton, NJ: Princeton University Press, 2005).

30 Tania Li, *The Will to Improve: Governmentality, Development and the Practice of Politics* (Chapel Hill, NC: Duke University Press, 2007).

31 Hugh Raffles, *In Amazonia: A Natural History* (Princeton, NJ: Princeton University Press, 2002), 7.

32 For example, see James McCarthy, "First World Political Ecology: Lessons from the Wise Use Movement," *Environment and Planning A*, 34 (2002): 1281–1302; W. Scott Prudham, "Poisoning the Well: Neo-liberalism and the Contamination of Municipal Water in Walkerton, Ontario," *Geoforum* 35, 3 (2004): 343–59; Bruce Braun, *The Intemperate Rainforest: Nature, Culture and Power on Canada's West Coast* (Minneapolis: University of Minnesota Press, 2002). On the intersection of neoliberalism and governmentality, see Nik Heynen, James McCarthy, Scott Prudham, and Paul Robbins, eds., *Neoliberal Environments: False Promises and Unnatural Consequences* (London and New York: Routledge, 2007).

33 Gavin Bridge, "Gas, and How to Get It," *Geoforum* 35, 4 (2004): 397.

34 Erich Zimmerman said as much eighty years ago with his great (though often misappropriated) axiom of resource geography, "Resources are not, they become." See Erich Zimmerman, *World Resources and Industries: A Functional Appraisal of the Availability of Agricultural and Industrial Materials* (New York: Harper, [1933] 1951).

35 For a prominent Canadian argument on the merits of merging environmental historical and political ecology, see Keeling and Sandlos, "Environmental Justice Goes Underground." A very useful recent historiography can be found in Liza Piper, "Knowing Nature through History," *History Compass* 11/12 (2013): 1139–49.

36 Keeling and Sandlos, "Environmental Justice Goes Underground," 124.

37 Piper, *Industrial Transformation*, 2.

38 Caroline Desbiens, *Power from the North: Territory, Identity, and the Culture of Hydroelectricity in Quebec* (Vancouver: UBC Press, 2013), 10, 13.

39 Emilie Cameron, *Far Off Metal River: Inuit Lands, Settler Stories, and the Making of the Contemporary Arctic* (Vancouver: UBC Press, 2015), xv; emphasis in original.

40 Nancy Lee Peluso, "What's Nature Got To Do With It? A Situated Historical Perspective on Socio-natural Commodities," *Development and Change* 43, 1 (January 2012): 79–104.

41 There have been a number of statements on historical political ecology. See, Christian Brannstrom, "What Kind of History for What Kind of Political Ecology?" *Historical Geography* 32 (2004): 71–88; Karl Offen, "Historical Political Ecology: An Introduction," *Historical Geography* 32 (2004): 7–18. Diana K. Davis, "Historical Approaches to Political Ecology," in the *Handbook of Political Ecology*, ed. Gavin Bridge, James McCarthy, and Tom Perrault (London: Routledge, 2015), 263–75. For an excellent example, see Dawn Day Biehler, "Permeable Homes: A Historical Political Ecology of Insects and Pesticides in US Public Housing," *Geoforum* 40, 6 (November 2009): 1014–23.

42 Scott Kirsch, *Proving Grounds: Project Plowshare and the Unrealized Dreams of Nuclear Earthmoving* (New Brunswick, NJ: Rutgers University Press, 2005); Susan Millar and Don Mitchell, "Spectacular Failure, Contested Success: the Project Chariot Bioenvironmental Programme," *Cultural Geographies* 5, 3 (1998): 287–302.

43 James C. Scott, *Seeing Like a State: How Certain Schemes to Improve the Human Condition Have Failed* (New Haven, CT: Yale University Press, 1998). This book also differs in scale;

human and environmental dislocation and disruption is lesser in the Stikine than in the cases described by Scott.

44 James Ferguson, *The Anti-Politics Machine: "Development," Depoliticization, and Bureaucratic Power in Lesotho* (Minneapolis: University of Minnesota Press, 1990), 252.

45 Ferguson, *Anti-Politics Machine,* 254.

46 Tania Murray Li, "Beyond 'the State' and Failed Schemes," *American Anthropologist* 107, 3 (2005): 384.

47 Jane Bennett, *Vibrant Matter: A Political Ecology of Things* (Chapel Hill, NC: Duke University Press, 2009), 94.

48 Michael Goldman, *Imperial Nature: The World Bank and Struggles for Social Justice in the Age of Globalization* (New Haven, CT: Yale University Press, 2005), 200. In Canada, see, for instance, Stunden Bower, *Wet Prairie.*

49 Richard White, *Railroaded: The Transcontinentals and the Making of Modern America* (New York, NY: W.W. Norton, 2011), xxi.

50 I am indebted to Tad McIllwraith for suggesting that I include an explicit statement to this effect.

51 Paul Tennant, *Aboriginal Peoples and Politics: The Indian Land Question in British Columbia, 1849–1989* (Vancouver: UBC Press, 1990).

52 The 1910 Tahltan Declaration, http://www.firstnations.de/media/05-3-declaration.pdf.

53 George T. Emmons, *The Tahltan Indians* (Philadelphia, PA: The University Museum, 1911); James Teit, "Notes on the Tahltan Indian of British Columbia," in *Boas Anniversary Volume: Anthropological Papers Written in Honor of Franz Boas on the Twenty-Fifth Anniversary of His Doctorate,* ed. Berthold Laufer (New York: G.E. Stechert and Company, 1906); James Teit, "Field Notes on the Tahltan and Kaska Indians, 1912–1915," *Anthropologica* 3 (1956): 39–213; James Teit, "Kaska Tales," *Journal of American Folklore* 30, 118 (1917): 427–73; James Teit, "On Tahltan (Athabaskan) Work, 1912," *Summary Reports of the Geological Survey of Canada,* Sessional Paper no. 26, 1912, 484–87; James Teit, "Tahltan Tales," *Journal of American Folklore* 32, 124 (1919): 198–250; James Teit, "Tahltan Tales," *Journal of American Folklore* 32, 124 (1919): 230; James Teit, "Two Tahltan Traditions," *Journal of American Folklore* 22, 85 (1909): 314–18; Sylvia L. Albright, *Tahltan Ethnoarchaeology* (Burnaby, BC: Department of Archaeology, Simon Fraser University, 1984), publication no. 15; Bruce B. MacLachlan, "Notes on Some Tahltan Oral Literature," *Anthropologica* 4 (1957): 1–9; David E. Friesen, *Aboriginal Settlement Patterns in the Upper Stikine River Drainage, Northwestern British Columbia* (master's thesis, University of Calgary, 1985); Judy Thompson, *Recording Their Story: James Teit and the Tahltan* (Seattle: University of Washington Press, 2007).

54 Cruikshank, *Do Glaciers Listen?,* 127–53.

55 Candis L. Callison, *A Digital Assemblage: Diagramming the Social Realities of the Stikine River Watershed* (master's thesis, MIT, 2002), 36.

56 Albright, *Tahltan Ethnoarchaeology,* 15–16.

57 Clarence Andrews, *Wrangell and the Gold of the Cassiar* (Seattle: Luke Tinker, 1937).

58 Davis, *The Sacred Headwaters,* 19–20.

59 Albright, *Tahltan Ethnoarchaeology;* Jan Krueger, Shane Conn, and Beth Moreau, "Tahl Tan Mission Study" (Opportunities for Youth Project, 1971), BCA, MS 2052. For a full history of early epidemic disease in British Columbia, see Robert Boyd, *The Coming Spirit of Pestilence: Introduced Infectious Diseases and Population Decline among Northwest Coast Indians, 1774–1874* (Vancouver: UBC Press, 2000).

60 Davis, *The Sacred Headwaters,* 20.

61 Edward R. Affleck, "Navigation on the Stikine River, 1862–1969," *British Columbia Historical News* 28, 1 (Winter 1994–95): 15–19; Art Downs, *Paddlewheels on the Frontier: The Story of British Columbia and Yukon Sternwheel Steamers* (Seattle: Superior, 1972).

62 Stewart Andrew Robb, "The Collins Overland or Russian Extension Telegraph Project" (master's thesis, Simon Fraser University, 1966).

63 Guy Lawrence, *Forty Years on the Yukon Telegraph* (Quesnel, BC: Caryall Books, 1965); Bill Miller, *Wires in the Wilderness: The Story of the Yukon Telegraph* (Victoria: Heritage House Publishing, 2004).

64 Guy Lawrence, "We Came to Get Rich – Damn Good Reason!" *Alaska Sportsman* (March 1943): 6–7, 22–28.

65 Jonathan Peyton, "Moving through the Margins: The "All-Canadian" Route to the Klondike and the Strange Experience of the Teslin Trail," in *Ice Blink: Narrating Northern Environmental History,* ed. Stephen Bocking and Brad Martin (Calgary, AB: University of Calgary Press, 2016); Andrews, *Wrangell and the Gold of the Cassiar;* Hamlin Garland, *Trail of the Goldseekers* (New York: MacMillan, 1899); Norman Lee, *Klondike Cattle Drive: The Journal of Norman Lee* (Vancouver, BC: Mitchell Press, 1960); W.D. MacBride, "From Montana to the Klondyke," *Caribou and Northwest Digest* 7 (April 1951) 8–9, 16a-19; (May 1951), 6–9, 19–28; (June 1951), 20–35; Richard Landry, "Telegraph Trail," *Alaska* 35, 11 (November 1969) and 36, 1 (January 1970); Lawrence, *Forty Years on the Yukon Telegraph;* Larry Pynn, *The Forgotten Trail: One Man's Adventures on the Canadian Route to the Klondike* (Toronto: Doubleday Canada, 1996).

66 Jonathan Peyton, "Imbricated Geographies of Conservation and Consumption in the Stikine Plateau," *Environment and History* 17, 4 (2011): 555–81; J.R. Bradley, *Hunting Big Game in Far Northwestern British Columbia* (New York: Mail and Express Job Print), 1904; Daniel J. Singer, *Big Game Fields of America, North and South* (London: Hodder and Stoughton, 1916); Frantz Rosenberg, *Big Game Shooting in British Columbia* (London: Hopkinson and Company, 1928); A. Bryan Williams, *Game Trails of British Columbia* (London: John Murray, 1925); Charles Alsopp Hindlip, 3rd Baron, "Hunting in British Columbia," *Travel and Exploration* 1, 3 (March 1909), 177–87; Pike, *Through the Subarctic Forest.*

67 William Cronon, "Kennecott Journey: The Paths In and Out of Town," in *Under an Open Sky: Rethinking America's Western Past,* ed. William Cronon, George Miles, and Jay Gitlin (New York: W.W. Norton, 1992), 28–51.

68 C.S. Nay, "Stream Geochemical Reconnaissance in the Canadian Cordillera" (Vancouver: Kennco Explorations (Western), n.d.).

69 Robert Gill et al., "Galore Creek Project, British Columbia: NI 43-101 Technical Report on Pre-Feasibility Study" (Vancouver: NovaGold Resources and AMEC Americas, 2011). http://www.novagold.com/upload/technical_reports/Galore_Creek_NI_43-101_2011_Final.pdf.

70 NovaGold, http://www.novagold.com/section.asp?pageid=22238.

Chapter 1: Cassiar, Asbestos

1 "Cassiar ... Do You Remember?" http://www.cassiar.ca/.

2 Conwest Exploration Company, "Annual Report 1951," 1–2. The samples were milled out at the Industrial Minerals Division of the federal Department of Mines and Technical Surveys.

3 Cassiar Asbestos Corporation, "Annual Report 1953," 2.
4 Cassiar Asbestos Corporation, "Annual Report 1952," 15; Cassiar, "Annual Report 1953," 2; Cassiar, "Annual Report 1954," 3.
5 Cassiar Asbestos Corporation, "Annual Report 1962," 10.
6 Suzanne Leblanc, *Cassiar: A Jewel in the Wilderness* (Prince George, BC: Caitlin Press, 2003), 116.
7 The types of asbestos are chrysotile, amosite, crocidolite, tremolite, actinolite, and anthophyllite.
8 George Mercer Dawson, *Report on an Exploration Made in the Yukon District, N.W.T. and Adjacent Portion of British Columbia, 1887* (Ottawa: Geological Survey of Canada, 1898). For a broad history of mining in northwest BC, see Douglas C. Baker, "Mining in Northern British Columbia," *Western Geography* 12 (2002): 1–12.
9 Cassiar Asbestos Corporation, "Annual Report 1952," 4.
10 The most ambitious study of asbestos mining in Canada is found in Jessica van Horssen, *A Town Called Asbestos: Environmental Contamination, Health, and Resilience in a Resource Community* (Vancouver: UBC Press, 2016).
11 Leblanc, *Cassiar,* 72, 79. For example, Leblanc claims that "by 1961 the 800 winter residents of the town included 135 married couples and 200 children."
12 "Employee Information Booklet," Cassiar Asbestos Corporation, n.d.
13 Leblanc, *Cassiar.*
14 "$400 Monthly Earned in Northern Post," *Ottawa Citizen,* December 31, 1952.
15 Leblanc, *Cassiar,* 71.
16 Leblanc, *Cassiar,* 75.
17 *Cassiar Courier* 1, 1 (November 1976), 1–2.
18 "Cassiar Output Starts in Fall," *Northern Miner,* July 31, 1952.
19 W.N. Plumb, "The Geology of the Cassiar Asbestos Deposit," June 1968, UNBC Archives, Cassiar Asbestos Collection, File 0806–003.
20 "History of the Stewart-Cassiar Highway," *Cassiar Courier* 4, 2 (December 1979), 2, 19.
21 G. Brett Maxwell, "A Brief Outline of the Economics and Northern Trucking Operations," Economics Staff Group, Northern Economic Development Branch, Department of Indian Affairs and Northern Development, February 18, 1971.
22 Maxwell, "A Brief Outline."
23 George Oswalt, "Yukon Dust," *Transport Times* 2, 6 (June 1971).
24 Cassiar Asbestos Corporation, "Transporting Asbestos Fibre Using the Cassiar-Stewart Highway," March 1978, UNBC Archives, Cassiar Asbestos Collection, File 2000.1–1204–017.
25 Cassiar Asbestos Corporation, "Annual Report 1957," 27.
26 Brinco, "Cassiar Annual Report 1981," 10.
27 Brian Pewsey, "Public Address," *Cassiar Reporter* 1 (October 1976), 1.
28 Gerry Doran, "Environmental – Cassiar" *Cassiar Reporter* 1 (October 1976), 4.
29 Ontario, "Report of the Royal Commission on Matters of Health and Safety Arising from the Use of Asbestos in Ontario" (Toronto: Ontario Ministry of the Attorney General, 1984). On risk and the framing of chrysotile asbestos as inert, see van Horssen, *A Town Called Asbestos,* 135–62.
30 Cassiar Asbestos Corporation, "A Submission to the Pollution Control Board of British Columbia for the Public Inquiry," January 1978.
31 Ibid.

32 Ibid.

33 Janet M. Landucci, "An Environmental Assessment of the Effects of Cassiar Asbestos Corporation on Clinton Creek, Yukon Territory," Department of Environment, Environmental Protection Service, Pacific Region (Regional Program Report no. 79–13), 1978.

34 Melvin S. Taylor, "The environmental control structure required to establish corporate policy, communications, corporate environmental management, and to meet all guidelines for a safe and healthy environment at Cassiar Asbestos Corporation Limited in Cassiar, BC" (Prepared for Cassiar Asbestos Corporation, April 1987).

35 Cassiar Asbestos Corporation, "Employee Information Booklet"; see also D.A. Enarson et al., "Respiratory Health in Chrysotile Asbestos Miners in British Columbia: A Longitudinal Study," *British Journal of Industrial Medicine* 45, 7 (1988): 459–63.

36 Leblanc, *Cassiar,* 141. Leblanc counts 6500 lawsuits filed against Cassiar between 1974 and 1991.

37 Cassiar Mining Corporation, "Submission of Cassiar Mining Corporation Regarding Proposed Occupational Safety and Health Regulation of Asbestos" (Prepared for the Workers' Compensation Board of British Columbia, January 1989).

38 There is no firm evidence of adverse health effects in the Cassiar case, but in most other places the health effects are profound. See van Horssen, *A Town Called Asbestos.*

39 Brinco Mining and Wright Engineers, *Pre-Feasibility Study, Volume 1: Technical and Cost Report,* December 1983.

40 A.A. Burgoyne, "Geology and Exploration, McDame asbestos deposit, Cassiar, BC," *The Canadian Mining and Metallurgical Bulletin* (May 1986); "Study Encouraging for Cassiar" *Whitehorse Star,* April 24, 1987.

41 Burgoyne, "Geology and Exploration."

42 Brinco, "Annual Report 1986," 6.

43 William Annett, "Shifting Fortunes," *BC Business* (February 1989), 28.

44 Leblanc, *Cassiar,* 154–63.

45 "Cassiar Completes First Shipment," *Northern Miner,* February 28–March 5, 2000, 2.

46 See www.discovery.ca/Shows/Jade-Fever.

47 John Bradbury and Michael Sendbuehler, "Restructuring Asbestos Mining in Western Canada," *Canadian Geographer* 32, 4 (1988): 296–306.

48 Annett, "Shifting Fortunes," 27.

49 Ibid., 27.

50 Princeton Mining Company, "Reorganization Plan – Summary," December 23, 1991. The company enumerated what would be lost in the event of closure under the title *What is at Stake?* Under "the Town of Cassiar," losses included "1050 residents; businesses, government offices; hospitals and schools; services all provided by Company including power, sewage, water, recreation, snow removal." Under "the McDame Project" was included "underground mine with more than 10 years remaining; 450 direct employees at Cassiar; 30 contract employees Stewart; 46 employees and contractors in Vancouver; $25 million per year in wages and benefits; $40 million per year to suppliers, materials and contractors: $75 million per year in export sales." Under "the Company" was included "Cassiar Mining Company head office in Vancouver; Open pit mine operated for 38 years; provincial taxes and payments of $5.7 million per year; directors – management and union representatives; international sales of products to 52 countries."

51 Bernice Trick, "Town Reduced to Boxes of Documents," *Prince George Citizen* June 9, 2000, 5.

52 The classic narrative of this type of corporate, bureaucratic safeguarding of ostensibly public records can be found in Peter Van Wyck, *The Highway of the Atom* (Montreal, QC: McGill-Queen's University Press, 2010). I am indebted to one of the reviewers for pointing out this source.

53 Elements of all of these are well represented in Suzanne Leblanc's exemplary community history, *Cassiar: A Jewel in the Wilderness*.

54 "In Memory," www.cassiar.ca/home/home.htm.

55 See www.facebook.com/album.php?aid=179307&id=170324924795.

56 Ramona Rose, "Ghost Town Turns Virtual Town: Preserving Cassiar's History Via Social Networking Sites," paper presented at the Canadian Historical Association Meeting, May 30, 2011, Fredericton, NB.

Chapter 2: Liberating Stranded Resources

1 Fortune Minerals, "Fortune Minerals announces railway development strategy for Mount Klappan metallurgical coal project," http://www.newswire.ca/news-releases/fortune-minerals-announces-railway-development-strategy-for-mount-klappanmetallurgical-coal-project-545248622.html.

2 United States, "Report of the Commission to Study the Proposed Alaska Highway to Alaska" (Department of State, Conference Series No. 14, Washington: Government Printing Office, 1933).

3 Joseph Gordon Smith, "British Columbia-Alaska Highway," *Pacific Travel Monthly* 1, 2 (1936), 10–15.

4 Ken Coates, *North to Alaska* (Anchorage: University of Alaska Press, 1992), 24–25.

5 Canada, British Columbia-Yukon-Alaska Highway Commission, "Report on Proposed Highway through British Columbia and the Yukon Territory to Alaska" (Ottawa: King's Printer, 1942 [report August 1941]).

6 Ken S. Coates and William Morrison, *Land of the Midnight Sun: A History of the Yukon* (Edmonton: Hurtig Publishers, 1988).

7 Coates, *North to Alaska*, 27–28.

8 Georgiana Ball, "Stikine History: Freighting to Watson Lake Aerodrome along the Stikine-Dease Corridor, 1941–43" (Telegraph Creek, BC: Stikine Community Association, 1992). Concern about possible Japanese invasion of Alaska likely played a role as well.

9 Ball, "Stikine History."

10 For transportation development in the years preceding Bennett's government, see John R. Wedley, "Laying the Golden Egg: The Coalition Government's Role in Post-War Northern Development," *BC Studies* 88 (Winter 1990–91): 58–92.

11 British Columbia Archives, GR 500, Box 16, File 1001–38, S.H. Bingham, "Report on Route and General Plan for a Railway in Northern British Columbia for the Wenner-Gren Development Co." (July 16, 1959). On the PNR, see Lawrence D. Taylor, "The Bennett Government's Pacific Northern Railway Project and the Development of British Columbia's 'Hinterland,'" *BC Studies* 175 (Autumn 2012): 35–56.

12 William E. Ryan, "BC Rail-to-Alaska Deal May Be Set Next Week: At Hush-Hush Conference," *Province*, August 29, 1959.

13 P.A. Gaglardi, "Brief on Transportation by British Columbia" (presented at Alaska-Yukon-British Columbia Conference, Victoria, July 20, 1960).

14 "Alaskan Road Plan Aids BC: $354 million project," *Province*, April 1, 1961.

15 "Cassiar Road Agreement Reached," *Whitehorse Star,* November 20, 1958; Philip Isard, "Northern Vision: Northern Development During the Diefenbaker Era" (master's thesis, Wilfred Laurier University, 2010).
16 "The Stewart-Cassiar Highway," *Cassiar Courier* (December 1979): 19; Thomas A. Elliott, "A Strategic Road to the North," *Western Business and Industry* (December 1964): 20–21.
17 Harper Reed Fonds, Box 1 and 3, BCA, MS 0516.
18 *Epigraph:* British Columbia, *Royal Commission on the British Columbia Railway,* vol. 1, 87.
19 Mark Crawford, "The Role of the State in the Economic Development of British Columbia: Case Studies of the Dease Lake Extension and Sukunka Coal," BC Project Working Paper (Victoria: BC Project, Political Science, University of Victoria, 1983), 29.
20 Ibid., 29.
21 Commissioners expressly avoided apportioning blame: "we are obliged to make inquiry into and concerning all aspects of the management and development of the railway. While these last-quoted words might be construed to mean that we should hold up to scrutiny the conduct of all past managers who helped shape the railway into its present form, and to condemn any whose performance we found wanting, we would be reluctant to make such an interpretation ... We have neither mandate nor inclination to join the hunt." British Columbia, *Royal Commission on the British Columbia Railway,* vol. 1.
22 There has been some suggestion of compromised impartiality among the three commissioners, who all had ties to the Social Credit party. Lloyd George McKenzie was counsel to the legislative committee that exonerated Gaglardi of previous corruption charges, David Chapman was the Social Credit nominee to the Canadian Anti-Inflation Board, and Syd Welsh was a well-known Social Credit party member.
23 W.A.C. Bennett, "Statement to the Royal Commission," University of British Columbia Archives (hereafter UBC Archives), BC Railway Commission Collection, Box 6, File 27, Exhibit 173A; cf. Morris Zaslow, *The Northward Expansion of Canada, 1914–1967* (McClelland and Stewart, 1988), 258.
24 W.A.C. Bennett, "Statement to the Royal Commission," UBC Archives, BC Railway Commission Collection, Box 6, File 27, Exhibit 173A.
25 Crawford, "Role of the State," 28–29.
26 Hedlin Menzies and Associates, "The Canadian Northwest Transportation Study, Final Report," prepared for the Government of Canada, Ministry of Transport (November 1970), ii.
27 BC Rail, "Dease Lake Rail Extension Study," in *Northern Extension: Fort St. James, Dease Lake and Fort Nelson: A Collection of Studies, 1961–1969* (Vancouver: British Columbia Railway, 1961–69), 10.
28 Bennett had many powerful allies in this claim. For instance, many Royal Commission submissions espouse the developmental necessity of the Dease Lake Extension. See submissions from the Kitimat-Stikine Regional District, the Mining Association of BC, Northern Development Council, and Cassiar Asbestos.
29 Touche and Ross and Company, "Assessment of the Dease Lake Extension," UBC Archives, BCR Collection, Box 7, File 38, Exhibit 264C, 1973.
30 British Columbia, Ministry of Recreation and Conservation, Fish and Wildlife Branch, "Submission" 16, UBC Archives, BCR Collection, Box 6, File 28, Exhibit 181.
31 Swan Wooster Engineering, "Report on Engineering Methods and Contract Administration," UBC Archives, BCR Collection, Box 8, File 1, Exhibit 264D, 1973.
32 British Columbia, *Royal Commission on the British Columbia Railway,* vol. 3, 106–7.

33 Nelson S. Hepburn, "British Columbia Railway: Appraisal of Engineering Aspects," UBC Archives, BCR Collection, Box 8, File 11, Exhibit 276; British Columbia, *Royal Commission on the British Columbia Railway,* vol 3, 168–69.
34 Stanley Oakes, "A Brief to the Royal Commission on the Affairs of the British Columbia Railway," UBC Archives, BCR, Box 5, File 16, Exhibit 60A, May 11, 1977.
35 British Columbia, *Royal Commission on the British Columbia Railway*, vol. 3.
36 BC Railway, "History of the Railway Line Commencing at Odell through to Dease Lake as of January 1975," UBC Archives, BCR Collection, Box 8, File 2, Exhibit 264H.
37 Swan Wooster Engineering, "Report on Engineering Methods and Contract Administration," UBC Archives, BCR Collection, Box 8, File 1, Exhibit 264D, 1973, i.
38 Swan Wooster Engineering, "Report on Engineering Methods and Contract Administration," UBC Archives, BCR Collection, Box 8, File 1, Exhibit 264D, 1973, ii.
39 British Columbia, Office of the Comptroller-General, "British Columbia Railway Company: Review of Effectiveness of Financial Systems," UBC Archives, BCR Collection, Box 7, File 37, Exhibit 264B, April 1973.
40 BC Railway, "History of the Railway Line Commencing at Odell through to Dease Lake as of January 1975," UBC Archives, BCR Collection, Box 8, File 2, Exhibit 264H.
41 Ibid.
42 Ibid.
43 British Columbia, *Royal Commission on the British Columbia Railway*, vol. 3, 101.
44 Canada experienced high inflation throughout the 1970s and into the 1980s, rising above 10 percent annually in 1974–75 and 1980–81.
45 British Columbia, *Royal Commission on the British Columbia Railway*, vol. 3, 81.
46 Swan Wooster Engineering, "Report on Engineering Methods and Contract Administration," UBC Archives, BCR Collection, Box 8, File 1, Exhibit 264D, 1973.
47 Fish and Wildlife Branch, BC Ministry of Recreation and Conservation, "Submission to the Royal Commission on BC Railway," UBC Archives, BCR Collection, Box 6, File 28, Exhibit 181, 16.
48 British Columbia Wildlife Federation, "Addendum Submission [L.S. Levernan Submission]," UBC Archives, BCR Collection, Box 6, File 30, Exhibit 189A.
49 David R. Bustard, *Environmental Problems Associated with the Abandonment of the Dease Lake Extension of BC Rail* (Smithers, BC: BC Fish and Wildlife Branch, 1977).
50 British Columbia Construction Association, "Brief Submitted to the Royal Commission on BC Railway," UBC Archives, BCR Collection, Box 5, File 23, Exhibit 69, May 13, 1977.
51 Crawford, "Role of the State, 26–27.
52 Ibid., 45.
53 Fish and Wildlife Branch, BC Ministry of Recreation and Conservation, "Submission to the Royal Commission on BC Railway," UBC Archives, BCR Collection, Box 6, File 28, Exhibit 181.
54 Fish and Wildlife Branch, BC Ministry of Recreation and Conservation, "Submission to the Royal Commission on BC Railway," UBC Archives, BCR Collection, Box 6, File 28, Exhibit 181, 9.
55 Bustard, "Environmental Problems."
56 Ibid. There is no copy of the photograph Bustard describes that can be suitably reproduced here.
57 Ibid. Again, the photograph in the file is not suitable for reproduction.

58 British Columbia Wildlife Federation, "Addendum Submission [L.S. Levernan Submission]," UBC Archives, BCR Collection, Box 6, File 30, Exhibit 189A.

59 Chinook Construction, "Brief," UBC Archives, BCR Collection, Box 5, File 6, Exhibit 32.

60 Fish and Wildlife Branch, BC Ministry of Recreation and Conservation, "Submission to the Royal Commission on BC Railway," UBC Archives, BCR Collection, Box 6, File 28, Exhibit 181, 9. There is a large literature on the riparian effects of forestry. See, for instance, Richard Rajala, "'Streams Being Ruined from a Salmon Producing Standpoint': Clearcutting, Fish Habitat, and Forest Regulation in British Columbia, 1900–45," *BC Studies* 176 (Winter 2012–13): 93–132.

61 Fish and Wildlife Branch, BC Ministry of Recreation and Conservation, "Submission to the Royal Commission on BC Railway," UBC Archives, BCR Collection, Box 6, File 28, Exhibit 181; Bustard, "Environmental Problems."

62 British Columbia Wildlife Federation, "Addendum Submission [L.S. Levernan Submission]," UBC Archives, BCR Collection, Box 6, File 30, Exhibit 189A.

63 Fish and Wildlife Branch, BC Ministry of Recreation and Conservation, "Submission to the Royal Commission on BC Railway," UBC Archives, BCR Collection, Box 6, File 28, Exhibit 181, 13.

64 Fish and Wildlife Branch, BC Ministry of Recreation and Conservation, "Submission to the Royal Commission on BC Railway," UBC Archives, BCR Collection, Box 6, File 28, Exhibit 181, 17.

65 Fish and Wildlife Branch, BC Ministry of Recreation and Conservation, "Submission to the Royal Commission on BC Railway," UBC Archives, BCR Collection, Box 6, File 28, Exhibit 181, 4.

66 Crawford, "Role of the State," 37.

67 British Columbia Wildlife Federation, "Submission," UBC Archives, BCR Collection, Box 6, File 29, Exhibit 189.

68 Stikine Copper, "Brief," UBC Archives, BCR Collection, Box 5, File 9, Exhibit 47.

69 Cassiar Asbestos Corporation, "Presentation," UBC Archives, BCR Collection, Box 5, File 52, Exhibit 126.

70 BC Railway, "History of the Railway Line Commencing at Odell through to Dease Lake as of January 1975," UBC Archives, BCR Collection, Box 8, File 2, Exhibit 264H, 28.

71 "Red Chris Mine: An Environmental Law Victory Can Still Be a Loss for the Environment," *West Coast Environmental Law Newsletter,* January 21, 2010, http://wcel.org/resources/environmental-law-alert/red-chris-mine-environmental-law-victory-can-still-be-loss-environ; Amanda Follett, "Feds Cutting Corners on Environmental Assessments: Supreme Court," *The Tyee,* January 21, 2010, http://thetyee.ca/Blogs/TheHook/Environment/2010/01/21/RedMine/.

72 Touche and Ross and Company, "Assessment of the Dease Lake Extension," UBC Archives, BCR Collection, Box 7, File 38, Exhibit 264C, 1973.

73 Jean-Paul Drolet, "The Demands and Limitations Governing the Mineral Production of Western Canada to 1990," paper presented at the Plenary Session on the Extraction of Our Mineral Resources (Vancouver, October 13, 1976).

74 Crawford, "Role of the State," 44; See also Envirocon and Pearse-Bowden Economic Consultants, *The Socioeconomic Effects of the BC Rail's Dease-Lake Extension on the Stuart-Trembleur Lakes Indian Band* (Vancouver: Report for the British Columbia Railway Company, February 1974).

75 BC Wildlife Federation, "Submission to the Royal Commission on BC Railway," UBC Archives, BCR Collection, Box 6, File 30, Exhibit 189, September 22, 1977.
76 British Columbia, *Royal Commission on the British Columbia Railway,* vol. 3, 78.
77 BC Railway, "Economic Report on Northern Extensions – Past, Present, Future," UBC Archives, BCR Collection, Box 11, File 20, February 28, 1975.
78 Alaska, Department of Commerce and Economic Development, "A Preliminary Study: Alaska-Canada Transcontinental Rail Connection to Contiguous United States," UBC Archives, BCR Collection, Box 6, File 6, Exhibit 138A, January 1977; Canalog Logistics, "The Alaska Rail Connection: A Review and Assessment of the Need for and Desirability of a Rail Connection to the Yukon and Alaska," September 1977, BCA, GR 500, Box 10, File 250.
79 Alaska Canada Rail Link, *Alaska Canada Rail Link Newsletter,* 1 (January 2006): 4.
80 Alaska Canada Rail Link, "Alternate Route Segment Assessment, Yukon and BC, Final Report," June 2006.
81 Alcan RailLink,"Rails to Resources to Ports: The Alaska Canada Rail Link Project, Phase 1 Feasibility Study, Executive Report," Whitehorse, March 2007.
82 By this point, Fortune had partnered with Posco Canada, a subsidiary of Korean conglomerate Posco. Under this equal partnership, the project would be known as the Arctos Anthracite Joint Venture.

Chapter 3: Corporate Ecology

1 Several dozen "power possibilities" documents were created between the mid-1940s and the mid-1960s. See, for instance, J. Doughty-Davies, *Power Possibilities of Nass and Stikine Rivers* (Victoria: Water Rights Branch, 1955); William Hick, "An Introduction to Northwestern British Columbia; (and) Major Underdeveloped Water Powers of Northern British Columbia, Geo. J. Smith," In *An Introduction to Northwestern British Columbia: Transactions of the Eight British Columbia Natural Resources Conference* (Victoria, n.p., 1955).
2 James C. Scott, *Seeing Like a State: How Certain Schemes to Improve the Human Condition Have Failed* (New Haven, CT: Yale University Press, 1999).
3 Matthew Evenden, *Fish Versus Power: An Environmental History of the Fraser River* (Cambridge, UK: Cambridge University Press, 2004); cf. Karl Boyd Brooks, *Public Power, Private Dams: The Hells Canyon High Dam Controversy* (Seattle: University of Washington Press, 2009).
4 Tina Loo, "Disturbing the Peace: Environmental Change and the Scales of Justice on a Northern River," *Environmental History* 12, 4 (2007): 895–919; Tina Loo and Meg Stanley, "An Environmental History of Progress: Damming the Peace and Columbia Rivers," *Canadian Historical Review* 92, 3 (Sept 2011): 399–427; cf. Nichole Dusyk, "Downstream Effects of a Hybrid Forum: The Case of the Site C Hydroelectric Dam in British Columbia, Canada," *Annals of the Association of American Geographers* 101, 4 (2011): 871–81; Matthew Evenden, ed., "Site C Forum," *BC Studies* 161 (2009): 93–114.
5 Caroline Desbiens, *Territory, Identity, and the Culture of Hydroelectrcity in Quebec* (Vancouver: UBC Press, 2013); cf. Hans Carlson, *Home Is the Hunter: The James Bay Cree and Their Land* (Vancouver: UBC Press, 2008); cf. John F. Hornig, ed., *Social and Environmental Impacts of the James Bay Hydroelectric Project* (Montreal and Kingston: McGill-Queen's University Press, 1999).

6 Joy Parr, *Sensing Changes: Technologies, Environments and the Everyday, 1953–2003* (Vancouver: UBC Press, 2009). For an ongoing example of the type of interactive work envisioned by Parr and her colleagues, see Joy Parr, Jessica van Horssen, and John van der Veen, *Megaprojects* (University of Western Ontario, 2008), http://megaprojects.uwo.ca.
7 Bruce Larsen, "Sun Writer Walks In on Crew: Secret Stikine Work Uncovered," *Vancouver Sun*, August 7, 1971, 1, 14. See also Moira Farrow, "BC Plans Five Hydro Dams: Three New Towns," *Vancouver Sun*, December 17, 1973. Brinco had actually been working in the area since 1966.
8 Andrea Demchuk, "The Stikine: Tahltans, Environmentalists, and BC Hydro." (master's thesis, Department of Political Science, University of British Columbia), 44–47.
9 BC Hydro (Hydroelectric Design Division), *Hydroelectric Development of the Stikine River Overview Study* (Victoria, January 1978). The purpose of the report was "to provide information for use in the project feasibility study and secondly, to provide information for the environmental impact analysis" (iv).
10 Demchuk, "The Stikine," 41–42. The reservoirs created by the dams would have been 103 kilometres and 25 kilometres in length, respectively.
11 BC Hydro, *Stikine-Iskut Hydroelectric Development: Progress report on Feasibility Studies* (Vancouver, December 1980).
12 L.G. Smith, "Taming BC Hydro: Site C and the Implementation of the BC *Utilities Commission Act*," *Environmental Management* 12, 4 (1990): 429–43.
13 Only the Revelstoke Dam (1984) was completed in this era, and it was planned and inaugurated before the creation of the Utilities Commission.
14 Jeremy Wilson, *Talk and Log: Wilderness Politics in British Columbia, 1965–96* (Vancouver: UBC Press, 1998); Frank Zelko, "Making Greenpeace: The Development of Direct Action Environmentalism in British Columbia," *BC Studies* 142/143 (2004): 241–77.
15 Full accounts of BC Hydro's Stikine-Iskut project can be found in Candis L. Callison, *A Digital Assemblage: Diagramming the Social Realities of the Stikine River Watershed* (master's thesis, MIT, 2002); Demchuk, *The Stikine;* and Lamont Bassett, "A River for the Taming: Megaplans in the Great Stikine Wilderness," *Harrowsmith* 8, 5 (February/March 1984): 32–45.
16 BC Hydro, *Prospectus: Northern Transmission Studies; Electrical Energy Transmission from Potential Hydroelectric Generation Projects on the Stikine/Iskut and Liard Rivers* (Vancouver, July 1980).
17 Gordon E. Beanlands and Peter N. Duinker, *An Ecological Framework for Environmental Impact Assessment in Canada* (Institute for Resource and Environmental Studies, Dalhousie University, Halifax, and Federal Environmental Assessment Review Office, Hull, QC, 1983); Robert B. Gibson, "From Wreck Cove to Voisey Bay: The Evolution of Federal Environmental Assessment in Canada," *Impact Assessment and Project Appraisal* 20, 3 (September 2002): 151–59; Bram Noble, *Introduction to Environmental Impact Assessment: A Guide to Principles and Practice* (Don Mills, ON: Oxford University Press, 2006).
18 Cited in John Faustmann, "The Future of the Stikine Basin," *Special Affairs Issue*, vol. 1, no. 1, (Vancouver: Public and Intergovernmental Relations Program, Indian Affairs, BC Region, 1982).
19 BC Hydro, "Program Designed to Care for Environment," *Northern Projects Journal* 1, 2 (May 1982): 5, supplement to the *Prince George Citizen*, May 28, 1982.

20 See Stephen Bocking, *Nature's Experts: Science, Politics and the Environment* (New Brunswick, NJ: Rutgers University Press, 2004).
21 For discussions on how knowledge and power are connected, see, in particular, Timothy Mitchell, *Rule of Experts: Egypt, Techno-politics, Modernity* (Berkeley: University of California Press, 2002); Anna Tsing, *Friction: An Ethnography of Global Connection* (Princeton, NJ: Princeton University Press, 2004).
22 For a discussion of non-knowledge and this process generally, see Jan Golinski, *Making Natural Knowledge: Constructivism and the History of Science* (Chicago: University of Chicago Press, 1998).
23 BC Hydro, *Preliminary Planning Report for the Stikine-Iskut Transmission System: A Summary of System Planning, Route Engineering, and Environmental Studies* (Vancouver, December 1982), 37.
24 BC Hydro, *Preliminary Planning,* 38.
25 For an outline of this critique, see David Demeritt, "Ecology, Objectivity, and Critique in Writings on Nature and Human Societies," *Journal of Historical Geography* 20, 1 (1994): 22–37. BC Hydro's budget and schedule did not plan for monitoring over long periods.
26 Scott, *Seeing Like a State,* 2.
27 BC Hydro, *Preliminary Planning,* 38.
28 B.R. Foster and E.Y. Rahs (Mar-Terr Enviro Research), "A Study of Canyon-Dwelling Mountain Goats in Relation to Proposed Hydro-electric Development in Northwestern British Columbia, Canada," *Biological Conservation* 33, 3 (1985): 212; this is a condensed version of B.R. Foster and E.Y. Rahs (Mar-Terr Enviro Research), *Relationship between Mountain Goat Ecology and Proposed Hydroelectric Development on the Stikine River, BC: Final Report on 1979–80 Field Studies, Prepared for BC Hydro (Generation Planning Department, System Engineering Division* (Vancouver, August 1981).
29 Michael Chabon, *Maps and Legends* (New York: McSweeney's, 2008).
30 BC Hydro, *Stikine-Iskut Rivers Hydroelectric Project Environmental Feasibility Studies; Economic Geology, Generation Planning Department – System Engineering Department* (Vancouver, December 1980). The "inventory" section was designed to "1. Describe and evaluate known and potential mineral ore bodies, coal deposits, aggregates, clays, limestones, quarry materials and other recoverable deposits within the study area. 2. Assess the potential for future petroleum and natural gas utilization within the study area." The "assessment" section was designed to "1. Identify and describe all minable ore bodies, deposits and/or oil and gas resources which would be lost as a result of hydroelectric development. 2. Identify and describe any ore bodies and deposits which may be exploited as a result of hydroelectric and related development of the study area. 3. Collaborate with Resources Economics Study in providing cost estimates of economic mineral, gas or oil deposits lost or foregone as a result of hydroelectric development. Estimate the economic benefit attainable from mineral, gas or oil utilization which may follow hydroelectric development in the area."
31 BC Hydro, *Stikine-Iskut Hydroelectric Development Feasibility Study: Hydrology, River Regime and Morphology* (Vancouver, October 1981).
32 Aresco, *Preliminary Archaeological Study of the Proposed Stikine-Iskut Hydroelectric Development, Draft Report* (Vancouver, April 1980). This report directly contradicts the claims made by the Tahltan about the importance of the river in their culture.
33 BC Hydro, *Stikine-Iskut Hydroelectric Development Exploration Program and Access Requirements* (Vancouver, December 1981), 6.

34 Ibid., 7–9.
35 Ibid., 13.
36 Ibid., 13.
37 These thirty organizations included both Stikine groups as well as the Telkwa Foundation, Sierra Club, Anglican Church of Canada, Southeast Alaska Conservation Council, Western Guide and Outfitters Association, Canadian Nature Foundation, Federation of BC Naturalists, BC Wildlife Federation, and Alaska Geographic Society. See Bentley Le Baron, *Stikine-Iskut Project: Social and Economic Impacts, Prepared for BC Hydro and Power Authority* (Vancouver, August 1982), 110–17.
38 In the north and interior of BC, direct action has proven costly and difficult to mobilize, and sometimes provokes mixed public response. For analyses of the complications related to direct-action political strategies in Canada, see Justin Page, *Tracking the Great Bear: How Environmentalists Recreated British Columbia's Coastal Rainforest* (Vancouver: UBC Press, 2014); and Ryan O'Connor, *The First Green Wave: Pollution Probe and the Origins of Environmental Activism in Ontario* (Vancouver: UBC Press, 2014).
39 Friends of the Stikine, *Newsletter* 1 (January 1981), 2.
40 Friends of the Stikine, *Newsletter* 1 (January 1981), 3–4.
41 Le Baron, *Stikine-Iskut Project,* 4.
42 Cited in Friends of the Stikine, *Newsletter* 1 (January 1981), 7. See also J. Plummer and R. Plummer, "Stikine Residents Speak: Locals Oppose Flooding," *Telkwa Foundation Newsletter* 3, 2 (Spring 1980), 6.
43 Friends of the Stikine, *Newsletter* 12 (November 1985), 4. Both groups submitted complementary briefs to the Pearse Commission on Pacific Fisheries, which placed land rights at the forefront of the issue. See C.R. Reitz, *Brief to Be Presented at the Pearse Commission Hearings, July 24–25, 1981,* Appendix 1 in Le Baron, *Stikine-Iskut Project.*
44 Pojar, "Transmission Links," 1.
45 Pojar, "Wildlife Impact," 3–4.
46 Cited in Friends of the Stikine, *Newsletter* 1 (January 1981), 7.
47 1910 Declaration of the Tahltan Tribe (also referred to as the 1910 Tahltan declaration), http://www.tndc.ca/pdfs/Tahltan%20Declaration.pdf.
48 Cited in Faustmann, "Future of the Stikine Basin," 13.
49 Ibid., 13.
50 Ibid., 14. The only specific study published from this program was Gregory Utzig et al. (Pedalogy Consultants), *Biogeoclimatic Zonation of the Stikine Basin, Prepared for the Association of United Tahltans* (November 1982).
51 Aspect Consultants and the Association of United Tahltans, *Stikine Basin Resource Analysis: An Evaluation of the Impacts of Proposed Hydroelectric Developments, Vol. 1* (Vancouver, 1981), 1.
52 Ibid., 2.
53 Ibid.
54 The prospectus was divided into nine sections that broadly reflected both potential economic opportunities and BC Hydro's assessment program: biophysical mapping, terrestrial wildlife, fisheries, agriculture, mineral resources, energy, wilderness recreation, tourism, socioeconomics, and the evaluation of land-use options.
55 The text of the analysis explains that "Tahltan resource people have been budgeted in the various study sectors as 'field assistants.' This should be recognized as a misnomer. In many situations it will be Tahltans who are directing the work of the moment, and the specialists

who are assisting. The term, however inappropriate, was used here for mere convenience" (Aspect Consultants, *Stikine Basin Resource Analysis,* 2).

56 British Columbia, Ministry of Sustainable Resource Management, "Cassiar Iskut-Stikine Land and Resource Management Plan," https://www.for.gov.bc.ca/tasb/slrp/lrmp/smithers/cassiar/plan/files/CIS-LRMP-November-2006.pdf.

57 Friends of the Stikine, *Newsletter* 1 (January 1981), 6; Faustmann, "Future of the Stikine Basin," 16. At a community meeting in Iskut in August 1980, one Tahltan Elder said that if a dam were built, he hoped to live long enough to be able to put some dynamite in it.

58 Le Baron, *Stikine-Iskut Project,* 3.

59 Ibid., 6.

60 BC Utilities Commission Hearings, *In the Matter of the Utilities Commission Act and in the Matter of an Application by BC Hydro and Power Authority Proceeding no. 48* (Vancouver, June 1, 1982), 94–97.

61 This is a common argument in most BC Hydro reports. See particularly Le Baron, *Stikine-Iskut Project;* and Ian Hayward and Associates, *Stikine/Iskut: Preliminary Environmental and Social Impact Assessment of the Stikine-Iskut Transmission System, Prepared for BC Hydro* (Vancouver, July 1982).

62 BC Hydro, *Stikine-Iskut Interim Report,* 1982.

63 P.J. McCart et al., *Stikine-Iskut Fisheries Studies 1979, Prepared for BC Hydro* (Vancouver, 1980), ix.

64 Richard Overstall, "Is There a Real Need for the Stikine-Iskut Dams?" *Telkwa Foundation Newsletter* 3, 2 (Spring 1980), 5; "Alaskans Fear Loss of Rich Estuary Habitat," *Telkwa Foundation Newsletter* 4, 2 (May-June 1981), 3.

65 BC Hydro, "Consultants Study Interplay of Stikine Estuary and River," *Northern Projects Journal* 1, 2 (May 1982), 1.

66 Florian Maurer, "Fledgling Fishery Threatened," *Telkwa Foundation Newsletter* 3, 2 (Spring 1980), 7.

67 For example, the Department of Fisheries and Oceans was concerned about the project but could allocate only $8,000 to the study of the issue.

68 Beak Consultants, *Preliminary Analysis of the Potential Impact of Hydroelectric Development of the Stikine River System on Biological Resources of the Stikine Estuary Prepared for BC Hydro* (Vancouver, September 1981).

69 Cited in Faustmann, "Future of the Stikine Basin," 12.

70 Ibid., 16.

Chapter 4: "Industry for the future"

1 Grant Thornton LLP, *Employment Impact Review,* prepared for the BC Ministry of Energy, Mines and Natural Gas (February 2013).

2 British Columbia, "Agreement Reached on Land to Build LNG Facility at Grassy Point," *BC Government Newsroom* (November 12, 2013). http://www.newsroom.gov.bc.ca/2013/11/agreement-reached-on-land-to-build-lng-facility-at-grassy-point.html.

3 See Nexen CNOOC, http://www.nexencnoocltd.com/en/Operations/ShaleGas/AuroraLNG.aspx.

4 Quoted in Shaun Thomas, "Aurora Signs an Exclusivity Deal for Grassy Point Development," *Northern View* (November 12, 2013), http://www.thenorthernview.com/news/231653041.html.

5 British Columbia, Ministry of Energy and Mines, "British Columbia's Natural Gas Strategy: Fuelling BC's Economy for the Next Decade and Beyond" (Victoria: Ministry of Energy and Mines, 2013), http://www.gov.bc.ca/ener/popt/down/natural_gas_strategy.pdf.
6 Quoted in Justine Hunter, "LNG's Greenhouse-Gas Impact Is Prompting Clark to Get Creative with Numbers," *Globe and Mail*, November 17, 2013, http://www.theglobeandmail.com/news/british-columbia/lngs-greenhouse-gas-impact-is-prompting-clark-to-get-creative-with-numbers/article15480455/.
7 See, for instance, J. David Hughes, *Drill, Baby, Drill: Can Unconventional Fuels Usher in a New Era of Energy Abundance?* (Santa Rosa, CA: Post Carbon Institute, 2013); J. David Hughes, "BC LNG: A Reality Check," *Watershed Sentinel*, January 17, 2014, http://watershedsentinel.ca/files/files/Hughes-BC-LNG-Jan2014.pdf; Mark Jaccard and Brad Griffin, "Shale Gas and Climate Targets: Can They Be Reconciled?" (Victoria: Pacific Institute for Climate Solutions, 2010). Karena Shaw and her collaborators have been at the forefront of research on the governance of new gas economies in northeast BC. See, in particular, Kathryn H. Garvie and Karena Shaw, "Oil and Gas Consultation and Shale Gas in British Columbia," *BC Studies* 184 (Winter 2014–15): 73–102; Kathryn H. Garvie and Karena Shaw, "Shale Gas Development and Community Response: Perspectives from Treaty 8 Territory, British Columbia," *Local Environment: The International Journal of Justice and Sustainability* (forthcoming).
8 Eleanor Stephenson, Alexander Doukas, and Karena Shaw, "Greenwashing Gas: Might a 'Transition Fuel' Label Legitimize Carbon-Intensive Natural Gas Development?" *Energy Policy* 46 (July 2012): 452–59.
9 Arn Keeling and John Sandlos, "Environmental Justice Goes Underground? Historical Notes from Canada's Northern Mining Frontier," *Environmental Justice* 2, 3 (2009): 117–25; John Sandlos and Arn Keeling, "Zombie Mines and the (Over)burden of History," *Solutions Journal* 4, 3 (June 2013): 80–83.
10 Nancy Lee Peluso, "What's Nature Got To Do With It? A Situated Historical Perspective on Socio-natural Commodities," *Development and Change* 43, 1 (2012): 79–104.
11 *Epigraph:* J.R. van der Linden, "The Western LNG Project," *Journal of Canadian Petroleum Technology* 22, 1 (January–February 1983): 55.
12 The body of literature on the history of energy in Canada is smaller than you might expect for a topic of such importance. See Paul Chastko, *Developing Alberta's Oil Sands: From Karl Clark to Kyoto* (Calgary, AB: University of Calgary Press, 2004); Paul Sabin, "Voices from the Hydrocarbon Frontier: Canada's Mackenzie Valley Pipeline Inquiry (1974–1977)," *Environmental History Review* 19, 1 (1995): 17–48; Erik Lizee, "Rhetoric and Reality: Albertans and Their Oil Industry under Peter Lougheed" (master's thesis, University of Alberta, 2010); Barry Ferguson, *Athabasca Oil Sands: Northern Resource Exploration, 1875–1951* (Regina: Canadian Plains Research Center, 1986); David Breen, *Alberta's Petroleum Industry and the Conservation Board* (Edmonton: University of Alberta Press, 1993); Larry Pratt and John Richards, *Prairie Capitalism: Power and Influence in the New West* (Toronto: McClelland and Stewart, 1979); Larry Pratt, *The Tar Sands: Syncrude and the Politics of Oil* (Edmonton: Hurtig, 1976). A new collection on energy history will add depth to these discussions: see Ruth Sandwell, ed., *Powering Up Canada: A Social History of Power, Fuel and Energy from 1600* (Montreal, QC: McGill-Queen's University Press, 2016).
13 David Nye, *Consuming Energy: A Social History of American Energies* (Cambridge, MA: MIT Press, 1998); David Nye, *Electrifying America: Social Meanings of a New Technology, 1880–1940* (Cambridge, MA: MIT Press, 1992).

14 Martin J. Pasqualetti, "Social Barriers to Renewable Energy Landscapes," *Geographical Review* 101, 2 (2011): 201–23; Martin J. Pasqualetti, "The Changing Energy Landscapes of North America," Energy and Environment Specialty Group (EESG) Plenary Lecture, 2013 AAG Annual Meeting, Los Angeles, California.

15 Gavin Bridge, "The Territorialities of Liquefied Natural Gas," paper presented at the 2013 AAG Annual Meeting, Los Angeles, California.

16 Gavin Bridge, "Gas and How to Get It," *Geoforum* 35, 4 (2004): 396.

17 International Energy Agency, *Golden Rules for the Golden Age of Gas: World Energy Outlook, Special Report on Unconventional Energy* (IEA: Paris, 2012).

18 Daniel Yergin and Michael Stoppard, "The Next Prize," *Foreign Affairs* 82, 6 (November–December 2003): 103–14.

19 Frederick Buell, "A Short History of Oil Cultures: Or, the Marriage of Catastrophe and Exuberance," *Journal of American Studies* 46, 2 (2012): 273–93.

20 *Epigraph:* van der Linden, "The Western LNG Project," 55.

21 There is an instructive parallel to be drawn with another ambitious extraction project on the eastern edge of the province at roughly the same time. At Tumbler Ridge, Denison Mines and the Teck Corporation were involved in complex negotiations with Japanese steel manufacturing partners that wished to secure fifteen-year agreements on coal extracted at the Quintette and Bullmoose mines. In fact, Teck partnered with Nissho Iwai at Bullmoose. The agreement was signed in 1981, but falling global coal prices and instability in the mining industry resulted in the forced renegotiation of the deal, almost from the outset. See W.H. Malkinson and Henry Wakabayashi, *North East Coal Development: Planning and Implementation* (Canadian Institute of Mining and Metallurgy, BC Ministry of Industry and Small Business Development, 1982); Paul Parker, "Canada-Japan Coal Trade: An Alternative Form of the Staple Production Model," *The Canadian Geographer* 41, 3 (1997): 248–67; Thomas Gunton, "Megaprojects and Regional Development: Pathologies in Project Planning," *Regional Studies* 37, 5 (2003): 505–19. Those interested in another related dimension of Japanese investment in northern Canada should see Larry Pratt and Ian Urquhart, *The Last Great Forest: Japanese Multinationals and Alberta's Northern Forests* (Edmonton: NeWest Press, 1994).

22 E.L. Forgues, "Western LNG Project," paper presented to Canadian Institute of Energy, November 1982, 18.

23 Anna Zalik, "Liquefied Natural Gas and Fossil Capitalism," *Monthly Review* 60, 6 (2008): 41–53.

24 Carter Energy, *Transpac Gas Project: Proposal, Liquefied Natural Gas (LNG) Export Project* (Vancouver: Carter Energy, 1981); Petro-Canada Exploration, Westcoast Transportation, and Mitsui and Company, *Rim Gas Project: Proposal to the Government of British Columbia* (Calgary: Petro-Canada Exploration, 1982).

25 British Columbia. "LNG in BC – Aurora LNG" (2014), http://engage.gov.bc.ca/lnginbc/lng-projects/aurora-lng/.

26 Albert Sigurdson, "British Columbia Gas Projects Could Mean New Industry in the Making," *Globe and Mail,* March 8, 1982, B14.

27 van der Linden, "The Western LNG Project," 55.

28 "Dome Signs Shipyard Pact Japanese," *Globe and Mail,* February 25, 1981, B15.

29 Arctic Pilot Gas Project; see also, Prince Rupert Regional Archives, 985–32 MS 672, Dome Petroleum, "Dome, LNG and British Columbia: Initial Perspectives on Dome Petroleum's Proposed Western Liquefided [sic] Natural Gas Project" (Calgary: Dome Petroleum, 1981).

30 R.H. McClelland, "Selection of Western LNG," ed. Hansard, British Columbia, Legislative Assembly 32nd Parl. 4th Sess 1982 Legislative Session (July 15, 1982). Govier was a former head of the Alberta Energy Resources Conservation Board (ERCB), the Alberta equivalent of the BCUC.

31 George W. Govier, *British Columbia, Commissioner Inquiry on British Columbia's Requirements, Supply and Surplus of Natural Gas and Natural Gas Liquids* (Victoria: Ministry of Energy, Mines and Petroleum Resources, 1982).

32 Dome Petroleum, "Western LNG Project Newsletter, Fifth Edition" (August 1982), 5–6; R.H. McClelland, *Project Selection: The Natural Gas Allocation Process, Statement by the Honourable R.H. McClelland Minister of Energy, Mines and Petroleum Resources* (Victoria: Ministry of Energy Mines and Petroleum Resources, 1982).

33 District of Kitimat, "Comparative Evaluation for Construction of a Liquefied Natural Gas Plant in Kitimat (Bish Cove), or Prince Rupert–Port Simpson (Grassy Point), BC (District of Kitimat, 1982).

34 Sid Tafler, "Struggle Over Site of $4 billion Plant Angers BC Towns," *Globe and Mail*, January 31, 1983, 10.

35 Forgues, "Western LNG Project," 18.

36 *Epigraph:* van der Linden, "The Western LNG Project," 58.

37 Dome Petroleum, "Western LNG Project Newsletter, Fifth Edition," 5.

38 Dome Petroleum, *Volume Three: Environmental Setting and Assessment for a Liquefied Natural Gas Terminal Grassy Point, Port Simpson Bay Northern British Columbia – Revision April 30, 1983* (Calgary: Dome Petroleum, 1981, 1983), 4.1.1.

39 Dome Petroleum, "Western LNG Project Newsletter, Fifth Edition," 6.

40 Clive Kessel, "Community Involvement in "Mega-Project" Planning: A Case Study of the Relationship between the Lax Kw'alaams Indian Band and Dome Petroleum" (master's thesis, University of British Columbia, 1984).

41 See Read Environmental and Planning Associates, *Impact Study Grassy Point LNG Project, Vol. 1–5* (prepared for the Port Simpson Indian Band, Vancouver, 1982).

42 Dome Petroleum and Lax Kw'alaams Indian Band Council, *Agreement* (Vancouver: Dome Petroleum, 1983).

43 Frank Cassidy and Norman Dale, *After Native Claims? The Claims of Comprehensive Claims Settlements for Natural Resources in BC* (Montreal, QC: Institute for Public Policy, 1988), 150–53; Harvey A. Feit, *Protecting Indigenous Hunters: The Social and Environmental Protection Regime in the James Bay and Northern Quebec Land Claims Agreement* (University of Michigan, Natural Resources Sociology Research Lab, 1982).

44 Cassidy and Dale, *After Native Claims?*, 153; Ralph Matthews and Nathan Young, "Development on the Margin: Development Orthodoxy and the Success of Lax Kw'alaams, British Columbia," *Journal of Aboriginal Economic Development* 4, 2 (Fall 2005): 100–8.

45 Dome Petroleum, *Environmental Setting and Assessment,* 4.0–1.

46 Ibid., 4.1–3, 4.1–14.

47 TERA Environmental Consultants, *Environmental Considerations in LNG Terminal Site Selection for Dome Petroleum Limited* (Calgary: Dome Petroleum, February 1981), 31.

48 Pan Canadian Consultants, *Environmental Overview of the Dome Petroleum Limited Western LNG Pipeline, a report prepared for Westcoast Transmission Company Limited and Dome Petroleum Limited* (Calgary: Dome Petroleum, 1981), 5.

49 Dome Petroleum, *Environmental Setting and Assessment,* 4.1–14a.

50 Canadian Resourcecon, *Social and Economic Impacts of the Western LNG Project* (Calgary: Dome Petroleum, 1982), 115.
51 Ibid., 118–22.
52 Dome Petroleum, *Environmental Setting and Assessment,* 4.9–1; on salvage ethnography on BC's North Coast, see Douglas Cole, *Captured Heritage: The Scramble for Northwest Coast Artifacts* (Vancouver: UBC Press, 1995).
53 Letter from Steve Acheson (research officer, Impact Assessment and Protection Section, Heritage Conservation Branch, Ministry of Provincial Secretary and Government Services) to Gary A. Webster (manager, environmental engineering, Dome Petroleum), April 30, 1982, cited in Dome, *Environmental Setting and Assessment,* i.
54 Ecology and Environment, *LNG Risk Analysis – Western LNG Project* (Calgary: Dome Petroleum, 1981).
55 Dome Petroleum, *Environmental Setting and Assessment,* 4.1–3b.
56 Ibid., 4.1–5.
57 Ibid., 4.1–6.
58 Ibid., 4.1–7.
59 *Epigraph:* van der Linden, "The Western LNG Project," 56.
60 "Dome's LNG Plan Criticized for Costs Exceeding Benefits," *Globe and Mail,* October 9, 1982, B16.
61 Paul Taylor, "Analysts Query Economics of Dome's LNG Scheme," *Globe and Mail,* February 1, 1983, B1.
62 David Stewart Patterson, "Dome Seeks Provincial Pacts to Beat Gas Export Deadline," *Globe and Mail,* February 2, 1983, B7.
63 Ibid., B7.
64 "Dome Talks on Asset Sale," *New York Times,* August 5, 1982; "Dome Might Leave Alsands," *New York Times,* February 5, 1982.
65 Cited in Jennifer Lewiston, "NEB Told Dome Canadian Content in LNG Plan Might Not Reach 50%," *Globe and Mail,* October 26, 1982, B1.
66 "Energy Price Drop Will Not Halt LNG Project" *Globe and Mail,* March 25, 1983, B3.
67 The British Thermal Unit (BTU) is a unit of measurement used predominantly in Dome's pricing (possibly intended to confuse the layperson). One million BTUs is equivalent to slightly more than a thousand cubic feet of gas.
68 Cited in Albert Sigurdson, "LNG Price to Japan Declines," *Globe and Mail,* April 26, 1983, B2.
69 "Chretien Feels Japan Reassured about LNG," *Globe and Mail,* April 19, 1983.
70 Cited in Paul Taylor, "Dome Tries to Save Japanese Gas Deal," *Globe and Mail,* January 13, 1984.
71 Cited in Albert Sigurdson, "Postponed LNG Delivery Called Disruptive to Buyers," *Globe and Mail,* October 21, 1983, B10.
72 Douglas Martin, "Dome's Fight to Untangle a Vast Financial Web," *New York Times,* August 19, 1984.
73 Cited in Kevin Cox, "Consortium Try to Salvage LNG Deal with Japanese Firms," *Globe and Mail,* March 19, 1985, B3.
74 Cited in "$3.5-Billion Liquefied-Natural-Gas Project Scrapped," *Montreal Gazette,* January 29, 1986, C3.
75 Buell, "A Short History of Oil Cultures," 273.

76 Andrew R. Thompson, British Columbia, "Statement of Proceedings: West Coast Oil Ports Inquiry" (Vancouver: West Coast Oil Ports Inquiry, 1978), 1.

77 Ibid., 98.

78 Ibid; Thomas R. Berger, *Northern Frontier, Northern Homeland: The Report of the Mackenzie Valley Pipeline Inquiry* (Ottawa: Minister of Supply and Services, 1977).

79 Canada, National Energy Board, *Report of the Joint Review Panel for the Enbridge Northern Gateway Project* (December 19, 2013), http://gatewaypanel.review-examen.gc.ca/clf-nsi/dcmnt/rcmndtnsrprt/rcmndtnsrprt-eng.html.

Chapter 5: Transmission

Epigraph: Quoted in Christopher Pollon, "Time to Get 'Wacky' Again: The Northwest Transmission Line," *The Tyee,* September 21, 2008.

1 See http://novagold.com/properties/galore_creek/overview/.

2 See http://www.imperialmetals.com/our-operations-and-projects/operations/red-chris-mine/overview.

3 See Mark Hume, "BC's Sacred Headwaters to Remain Protected from Drilling," *Globe and Mail,* December 18, 2012, http://www.theglobeandmail.com/news/british-columbia/bcs-sacred-headwaters-to-remain-protected-from-drilling/article6504385/.

4 Under the terms of the sale negotiation, Fortune retains the right to repurchase Arctos Anthracite for the next ten years; see http://www.fortuneminerals.com/news/press-releases/press-release-details/2015/Fortune-Minerals-and-POSCAN-complete-sale-of-Arctos-coal-licenses-to-BC-Rail-with-a-10-year-repurchase-option/default.aspx.

5 For example, after the Red Chris Supreme Court ruling (discussed below), BC Environment Minister Barry Penner complained to the *Globe and Mail* that the process was "cumbersome and costly," with built-in duplications that held up investment. See Mark Hume, "BC Mine Falling Through Very Large Crack in System," *Globe and Mail,* February 1, 2010, A8.

6 Meinhard Doelle and A. John Sinclair, "Time for a New Approach to Public Participation in EA: Promoting Cooperation and Consensus for Sustainability," *Environmental Impact Assessment Review* 26, 2 (2006): 185–205; A. John Sinclair and Meinhard Doelle, "Using Law as a Tool to Ensure Meaningful Public Participation in Environmental Assessment," *Journal of Environmental Law and Practice* 12, 1 (2003): 27–53; Robert Gibson et al., "Strengthening Strategic Environmental Assessment in Canada: An Evaluation of Three Basic Options," *Journal of Environmental Law and Practice* 20, 3 (2010): 175–211; Ciaran O'Faircheallaigh, "Public Participation and Environmental Impact Assessment: Purposes, Implications, and Lessons for Public Policy Making," *Environmental Impact Assessment Review* 30, 1 (2010): 19–27.

7 Pollon, "Time to Get 'Wacky.'"

8 Mining Association of British Columbia, *Report on the Electrification of the Highway 37 Corridor: A discussion on the potential benefits of a new power transmission line to northwest British Columbia* (September 2008), http://www.mining.bc.ca/sites/default/files/resources/final_mabcreport_electrification_of_highway_37_oct1.pdf.

9 Wendy Stueck, "New Power Transmission Line Could Spur Mining Development," *Globe and Mail,* October 2, 2007, S3.

10 Office of the Prime Minister of Canada, "PM Announces Canada's Investment in Northwest Transmission Line," Office of the Prime Minister press release, September 16, 2009; cited

in, "Government of Canada Annouces $130 Million Investment in Northwest Transmission Line," *Northwest Development Inititaive Trust,* September 16, 2009, http://www.northerndevelopment.bc.ca/news/government-of-canada-announces-130-million-investment-in-northwest-transmission-line/.

11 Justine Hunter, "Harper Pledges $130 Million for Northern BC Power Line," *Globe and Mail,* September 16, 2009, http://www.theglobeandmail.com/news/national/british-columbia/harper-pledges-130-million-for-northern-bc-power-line/article1290208/; Mark Hume, "Does BC's New Power Line Fall Short of Being Green?" *Globe and Mail,* September 21, 2009, http://www.theglobeandmail.com/news/national/does-bcs-new-power-line-fall-short-of-being-green/article1295110/; Pollon, "Time to Get 'Wacky'"; Christopher Pollon, "Gordon Campbell's $400 Million Power Line Bet," *The Tyee,* November 13, 2008, http://thetyee.ca/News/2008/11/13/Hwy37/.

12 Scott Simpson, "Collapse of Galore Mine Project Leaves Transmission Line in Limbo," *Vancouver Sun,* November 28, 2007, http://www.canada.com/vancouversun/news/business/story.html?id+12ec5fd0-604c-4758-83d6-685972efdf8.

13 The Northern Development Initiative Trust describes itself as a "grass roots assembly of forty communities and forty-nine other groups and companies, including First Nations, mining and power companies, equipment suppliers, contracting and engineering companies, mining industry associations and First Nations joint venture companies." See Janine North, "Northern BC Communities Electrified Over Federal $130 Million Investment in the Northwest Transmission Line," *Northern Development Initiative Trust,* September 17, 2009, http://www.northerndevelopment.bc.ca/news/northern-bc-communities-electrified-over-federal-130-million-investment-in-the-northwest-transmission-line/.

14 "Corporate Donors Ask BC Liberals for Half-Billion-Dollar Handout," *Dogwood Initiative Bulletin,* February 20, 2007.

15 "Power Line Fate in Hands of Government," *Terrace Standard,* January 12, 2011, http://www.bclocalnews.com/bc_north/terracestandard/news/113249979.html; British Columbia, Ministry of the Environment, "Northwest Transmission Line Approved," BC Gov. Information Bulletin, February 23, 2011. https://www.bchydro.com/content/dam/hydro/medialib/internet/documents/projects/ntl/NTL_info_bceao_approved_feb2011.pdf.

16 Christopher Pollon, "Reinvent Environmental Assessment in BC, Say Critics," *The Tyee,* Novermber 8, 2012. http://thetyee.ca/News/2012/11/08/Reinvent-Environmental-Assessment-in-BC/.

17 BC Hydro, "Northwest Transmission Line Agreement Will Create Jobs, Power BC's Northwest" (May 28, 2010), https://www.bchydro.com/news/press_centre/news_releases/2010/ntl_agreement_will_create_jobs_power_bc_nw.html.

18 BC Environmental Assessment Office, http://www.eao.gov.bc.ca/.

19 A signal example of this type of work can be found in Mara J. Goldman, Paul Nadasdy, and Matthew D. Turner, *Knowing Nature: Conversations at the Intersection of Political Ecology and Science Studies* (Chicago: University of Chicago Press, 2011).

20 There is a very large literature dealing with the intersection of science and politics. See Stephen Bocking, *Nature's Experts: Science, Politics, and the Environment* (New Brunswick, NJ: Rutgers University Press, 2004); Sheila Jasanoff, *Designs on Nature: Science and Democracy in Europe and the United States* (Princeton, NJ: Princeton University Press, 2005); Sheila Jasanoff, *The Fifth Branch: Science Advisors as Policy Makers* (Cambridge, MA: Cambridge University Press, 1990); Tim Forsyth, *Critical Political Ecology: The Politics of*

Environmental Science (London: Routledge, 2003); Peter J. Usher, "Traditional Ecological Knowledge in Environmental Assessment and Management," *Arctic* 53, 2 (June 2000): 183–93; Paul Nadasdy, *Hunters and Bureaucrats: Power, Knowledge and Aboriginal-State Relations in the Southwest Yukon* (Vancouver: UBC Press, 2004).

21 British Columbia Transmission Corporation (hereafter BCTC), *Northwest Transmission Line Project, Application for an Environmental Assessment Certificate, Part 1* (Vancouver: Rescan Environmental Services [Project #832–2] January 2010), xvi; Gordon E. Beanlands and Peter N. Duinker. "An Ecological Framework for Environmental Impact Assessment," *Journal of Environmental Management* 18 (1984): 267–77.

22 BCTC, *Northwest Transmission Line,* xvi. VECs were selected by BCTC in consultation with local interest groups, including First Nations.

23 Bocking, *Nature's Experts.*

24 BCTC, *Northwest Transmission Line,* xvii.

25 Stephen J. Genuis, "Fielding a Current Idea: Exploring the Public Health Impact of Electromagnetic Radiation," *Public Health* 122, 2 (February 2008): 113–124; Anders Ahlbom and Maria Feychting, "Electromagnetic Radiation: Environmental Pollution and Health," *British Medical Bulletin* 68, 1 (2008): 157–65.

26 BCTC, *Northwest Transmission Line,* xxvi.

27 Ibid., xviii.

28 Ibid., xxv.

29 Ibid., xxv.

30 Ibid., xxv.

31 Ibid., xxix.

32 Ibid., xxix–xxx.

33 Ibid., xvii.

34 These are identified as "the Eskay Creek Mine; the Forrest Kerr power project; the Red Chris copper and gold mine project; the Galore Creek copper and gold mine project; forestry activities, such as Kalum, Nass and Kispiox Timber Supply Area; the communities of Terrace, Stewart, Gatwangak, Gitanyow, Rosswood, New Aiyansh, Gitwinksihlkw, Laxgalts'ap, and Gingolx; roads and vehicle traffic; provincial parks and tourism; the existing 1L381 and 1L387 transmission lines." BCTC, *Northwest Transmission Line,* xvii.

35 Ibid., xvii.

36 Bram Noble, "Cumulative Environmental Effects and the Tyranny of Small Decisions: Towards Meaningful Cumulative Effects Assessment and Management," *Natural Resources and Environmental Studies Institute Occasional Paper* 8 (UNBC, Prince George, BC, 2010); see also Peter N. Duinker and Lorne A. Greig, "The Impotence of Cumulative Effects Assessment in Canada: Ailments and Ideas for Redeployment," *Environmental Management* 37, 2 (2006): 153–61.

37 The BCTC identifies "the seven First Nations" with potential interest in the NTL as Kitselas First Nation, Kitsumkalum band, Metlakatla First Nation, Lax Kw'alaams, Gitxsan Hereditary Chiefs, Gitanyow Hereditary Chiefs, and Tahltan Nations (as represented by the Tahltan Central Council).

38 *Haida Nation v. British Columbia (Minister of Forests),* [2004] 3 S.C.R. 511, 2004 SCC 73. *Taku River Tlingit First Nation v. British Columbia (Project Assessment Director),* [2004] 3 S.C.R. 550, 2004 SCC 74. For an analysis of these two cases and their impacts on Aboriginal law and environmental law, see Thomas Isaac, Tony Knox, and Sarah Bird, "The

Crown's Duty to Consult and Accommodate Aboriginal Peoples: The Supreme Court of Canada's Decisions in *Haida Nation v. BC* and *Weyerhaeuser and Taku River Tlingit First Nation v. BC,*" *Aboriginal Law Group* (2005): 1–8.

39 Morin has since taken a leave from THREAT to work on the BCEAO file on the Prosperity Mine (Taseko Mines) in the Chilcotin. Prosperity planned to turn Fish Lake into a tailings pond. It was the first major proposal the BCEAO has turned down, though the company has since brought a suit against the federal government in the BC Supreme Court for breach of legal duties. THREAT employs Tahltan expertise to direct industry and government on how to better protect the heritage, culture, and resources of the Tahltan Nation. THREAT provides independent technical expertise in the environmental assessment process, in order to ensure that Tahltan values are addressed.

40 BC Environmental Assessment Office (hereafter BCEAO), 2009 Aboriginal (A9) Comments on the NTL Project Draft Terms of Reference, 16–17, http://a100.gov.bc.ca/appsdata/epic/documents/p299/1260404358142_5afbeac073c07828125ceed6a71cf5e955b8b3bf9ffc43dbd2f6cf4b9dd327df.pdf.

41 BCEAO, 2009 Aboriginal (A9) Comments on the NTL Project Draft Terms of Reference, 17, http://a100.gov.bc.ca/appsdata/epic/documents/p299/1260404358142_5afbeac073c07828125ceed6a71cf5e955b8b3bf9ffc43dbd2f6cf4b9dd327df.pdf. Details of this concern include, but are not limited to, historical use and occupation of the land, summary of traditional use areas and resources, community values, value of fish and wildlife, access to traditional knowledge, demographic indicators, social, cultural and community issues and services, etc.

42 BCEAO, 2009 Aboriginal (A9) Comments on the NTL Project Draft Terms of Reference, 17, http://a100.gov.bc.ca/appsdata/epic/documents/p299/1260404358142_5afbeac073c07828125ceed6a71cf5e955b8b3bf9ffc43dbd2f6cf4b9dd327df.pdf.

43 BCEAO, 2009 Aboriginal (A9) Comments on the NTL Project Draft Terms of Reference, 17, http://a100.gov.bc.ca/appsdata/epic/documents/p299/1260404358142_5afbeac073c07828125ceed6a71cf5e955b8b3bf9ffc43dbd2f6cf4b9dd327df.pdf.

44 BCEAO, 2009 Aboriginal (A9) Comments on the NTL Project Draft Terms of Reference, 22, http://a100.gov.bc.ca/appsdata/epic/documents/p299/1260404358142_5afbeac073c07828125ceed6a71cf5e955b8b3bf9ffc43dbd2f6cf4b9dd327df.pdf.

45 BCEAO, 2009 Aboriginal (A9) Comments on the NTL Project Draft Terms of Reference, 22, http://a100.gov.bc.ca/appsdata/epic/documents/p299/1260404358142_5afbeac073c07828125ceed6a71cf5e955b8b3bf9ffc43dbd2f6cf4b9dd327df.pdf.

46 See Courtney Fidler, "Aboriginal Participation in Mineral Development: Environmental Assessment and Impact Benefit Agreements" (master's thesis, University of British Columbia, 2008).

47 Golder Associates, "Northwest Transmission Line: Environmental Assessment Certificate Application Review First Nations Consultation Summary" (prepared for BC Hydro Aboriginal Relations and Negotiations [Report no. 08–1477–0016], October 2010).

48 BCEAO, 2009 Public (P9) Comments on the NTL Project Draft Terms of Reference, 6, http://a100.gov.bc.ca/appsdata/epic/documents/p299/1260404496735_5afbeac073c07828125ceed6a71cf5e955b8b3bf9ffc43dbd2f6cf4b9dd327df.pdf.

49 BCEAO, 2009 Public (P9) Comments on the NTL Project Draft Terms of Reference, 12–13, http://a100.gov.bc.ca/appsdata/epic/documents/p299/1260404496735_5afbeac073c07828125ceed6a71cf5e955b8b3bf9ffc43dbd2f6cf4b9dd327df.pdf.

50 BCEAO, 2009 Public (P9) Comments on the NTL Project Draft Terms of Reference, http://a100.gov.bc.ca/appsdata/epic/documents/p299/1260404496735_5afbeac073c07828125ceed6a71cf5e955b8b3bf9ffc43dbd2f6cf4b9dd327df.pdf.
51 Alaska–Canada Energy Coalition: http://www.acecoalition.com.
52 Scott Simpson, "BC Power Line Plans Have Alaskans Buzzing," *Vancouver Sun,* January 13, 2010, http://www.canada.com/story.html?id=3c68bd0a-73dc-4d20-9245-a0ff468fd5f5.
53 The Northwest Power Line Coalition produced a short promotional video that can be seen online at https://www.youtube.com/watch?v=kVe915Cykyw.
54 North, "Northern BC Communities."
55 Tahltan First Nation and International Institute of Sustainable Development (IISD), *Out of Respect: The Tahltan, Mining, and the Seven Questions to Sustainability* (Report of the Tahltan Mining Symposium, April 4–6, 2003, Dease Lake, BC).
56 A.F. Buckham and B.A. Latour, *The Groundhog Coalfield, British Columbia, Bulletin 16 of Geological Survey of Canada* (E. Cloutier: King's Printer, 1950); R.C. Campbell-Johnston, "The Importance of Groundhog Coal," *British Columbia Magazine* 9, 10 (October 1913): 571–76; G.S. Malloch, Groundhog Coal Field: Geological Survey of Canada A Series Maps: 106A (Geological Survey of Canada Memoir 69, 1913).
57 Fidler, "Aboriginal Participation," 42.
58 Northwest Power Line Coalition, *Highway 37 Report,* May 2009.
59 Robert Gill, et al., "Galore Creek Project, British Columbia: NI 43-101 Technical Report on Pre-Feasibility Study." Vancouver: NovaGold Resources and AMEC Americas, 2011. http://www.novagold.com/_resources/projects/technical_report_galore_creek.pdf.
60 Stone Sheep (sometimes called Stone's Sheep) are named for Andrew J. Stone, hunter-naturalist with the American Museum of Natural History. Stone "discovered" a black sheep (subsequently called the Stone Sheep) that was different from the Dall sheep, its more northerly counterpart. See Andrew J. Stone, *Journals of Andrew J. Stone: Expeditions of Arctic and Subarctic America after Wild Sheep, Grizzly, Caribou, and Muskoxen,* ed. Margaret R. Frisina (Long Beach, CA: Safari Press, 2010); British Columbia, Ministry of Environment, Land and Parks, BC Parks Division, *Stikine Country Protected Areas Technical Background Information Summary, Draft Copy* (July 31, 2000).
61 Red Chris also lies close to the properties owned by environmental advocates Wade Davis and David Suzuki, who have both been instrumental in publicizing the threat to the Sacred Headwaters.
62 Christopher Pollon, "Report from the Edge of BC's Copper Rush," *The Tyee,* January 13, 2011, http://thetyee.ca/News/2011/01/13/Stikine/.
63 "Red Chris Mine: An Environmental Law Victory Can Still Be a Loss for the Environment," *West Coast Environmental Law Newsletter,* January 21, 2010, http://wcel.org/resources/environmental-law-alert/red-chris-mine-environmental-law-victory-can-still-be-loss-environ; Amanda Follett, "Feds Cutting Corners on Environmental Assessments: Supreme Court," *The Tyee,* January 21, 2010, http://thetyee.ca/Blogs/TheHook/Environment/2010/01/21/RedMine/.
64 Wade Davis, "A Power Line to Nowhere: What Coal Mining Means to British Columbia's Sacred Headwaters," *The Walrus* December 18, 2013, http://thewalrus.ca/a-power-line-to-nowhere/. Davis shows that the agreement between Red Chris and BC Hydro requires BC Hydro to buy the line from the mining company upon its completion for $52 million dollars.

65 Pollon, "Report from the Edge of BC's Copper Rush."
66 "Fortune Minerals and Tahltan Nation Enter into Environmental Assessment Cooperation Agreement for the Mount Klappan Anthracite Coal Project, British Columbia," *Canada NewsWire,* February 9, 2009, https://business.highbeam.com/1758/article-1G1-193303982/fortune-minerals-tahltan-nation-enter-into-environmental.
67 This extraordinary event is chronicled in Monte Paulson, "A Gentle Revolution," *The Walrus* 2, 10 (December 2005/January 2006): 64–75.
68 "Shell's Suspension in Klappan Extended for Another Year," *Terrace Standard,* February 15, 2011, http://www.terracestandard.com/news/116244339.html.
69 http://www.sacredheadwaters.com/; http://www.firstnations.de/mining/tahltan-klabona.htm.
70 Tyler Allan McCreary, "Struggles of the Tahltan Nation," *Canadian Dimension,* November 1, 2005, 14; Mark Hume and Wendy Stueck, "Elders Stage Month-Long Sit In Waiting to Speak to Chief," *Globe and Mail,* February 26, 2005, A5.
71 See http://www.firstnations.de/media/05-3-klappan.pdf.
72 Wade Davis, *The Sacred Headwaters: The Fight to Save the Stikine, Skeena, and Nass* (Vancouver: Greystone, 2011).
73 GW Solutions, "Coal Bed Methane and Salmon: Assessing the Risks," (prepared for the Pembina Institute, Calgary, May 2008). Serious groundwater contamination has been detected as a result of similar fracking technology in upstate New York and northeastern Pennsylvania, in the Utica and Marcellus shale formations, respectively. The literature on fracking is significant. See, for instance, Stephen G. Osborn et al., "Methane Contamination of Drinking Water Accompanying Gas-Well Drilling and Hydraulic Fracturing," *PNAS* 108, 19 (May 10, 2011): 1–5.
74 Paul R. Josephson, *Indistrialized Nature: Brute Force Technology and Transformation of the Natural World* (Washington, DC: Island Press, 2002).
75 Northwest Power Line Coalition, *Highway 37 Report,* May 2009.
76 *Epigraph:* Cited in Christopher Pollon, "BC's Tahltan and the Road to Power," *The Tyee,* January 14, 2011, http://thetyee.ca/News/2011/01/14/TahltanRoadToPower/.
77 Evidence is mostly anecdotal from my time spent in the Stikine, but see Christopher Pollon, "Northwest Power Line Grows, So Does Controversy," *The Tyee,* July 18, 2011, http://thetyee.ca/News/2011/07/18/NorthwestTransmissionLine/.
78 Pollon, "Report from the Edge of BC's Copper Rush."
79 See http://governmentcaucus.bc.ca/mikemorris/wp-content/uploads/sites/45/2014/02/Tahltan-Nation-Partnerships-in-Energy.pdf.

Conclusion

Epigraph: Edward Hoagland, *Notes from the Century Before: A Journal From British Columbia* (Toronto: Random House Canada, 1969), 15.

1 Edward Hoagland, *Notes from the Century Before: A Journal From British Columbia* (New York: Random House, 1969), 15.
2 Frederick Jackson Turner, *The Frontier in American History* (New York: Holt, 1921); Theodore Roosevelt, *The Winning of the West,* vols. 1–4 (New York: Putnam and Sons, 1889–1896).
3 Harold Innis, *The Fur Trade in Canada: An Introduction to Canadian Economic History* (New Haven, CT: Yale University Press, 1930); J.M.S. Careless, *Canada: A Story of Challenge*

(Toronto: Macmillan, 1953); cf. Donald Creighton, *The Commercial Empire of the St. Lawrence 1760–1850* (Toronto: Ryerson Press, 1937).

4 R.M. Patterson, *Trail to the Interior* (Toronto: Macmillan, 1966), xii.

5 See Edward Hoagland, *Early in the Season: A British Columbia Journal* (Vancouver: Douglas and McIntyre, 2009).

6 For an analysis of the frontier theory and its potential application in the environmental history of the Canadian north, see Peter R. Mulvihill, Douglas C. Baker, and William R. Morrison, "A Conceptual Framework for Environmental History in Canada's North," *Environmental History* 6, 4 (October 2001): 611–26.

7 Ann Laura Stoler, *Along the Archival Grain: Epistemic Anxieties and Colonial Common Sense* (New Haven, CT: Princeton University Press, 2010).

8 Quoted in "BC's Mining Sector Pleased Northwest Transmission Line Approved," *Northern BC Business,* February 25, 2011, http://northernbcbusiness.com/2011/02/27/bcs-mining-sector-pleased-northwest-transmission-line-approved/.

9 Craig Wong, "Miners Set to Benefit from New Power Line to Be Built in Northwestern BC," *Winnipeg Free Press,* May 15, 2011, http://agoracom.com/ir/CopperFoxMetals/forums/discussion/topics/484112-miners-set-to-benefit-from-ntl/messages/1553784.

10 Christopher Pollon, "Alaska Power and the Bleeding of the Northwest," *The Tyee,* February 18, 2011, http://thetyee.ca/News/2011/02/18/AlaskaPower/.

11 For information on these projects, see Altagas, http://www.altagas.ca/.

12 Thomas McIllwraith, *We Are Still Didene: Stories of Hunting and History in Northern BC* (Toronto: University of Toronto Press, 2012).

13 British Columbia, "Factsheet: LNG Project Proposals in British Columbia," March 7, 2016, https://news.gov.bc.ca/factsheets/factsheet-lng-project-proposals-in-british-columbia.

14 *Tsilhqot'in Nation v. British Columbia,* [2014] SCC 44.

15 Chad Norman Day, "Update from Chad Day about Mount Polley Mine," *Tahltan Central Council,* August 6, 2014, http://tahltan.ca/update-chad-day-mount-polley-mine/.

16 Hoagland, *Notes from the Century Before,* 15.

Bibliography

Archival Sources

British Columbia Archives and Records Services, Victoria

GR0500, Commission on British Columbia Railway (1977) Records, 1977–78.

GR0709, Records Relating to BC Department of Highways Projects, 1926–71.

GR1095, BC Department of Mines, Gold Commissioner Records, 1898.

MS0098, George Ball Fonds, 1927–29.

MS0516, T.F. Harper Reed Fonds, 1918–75.

MS2052, Jan Krueger, Shane Conn, and Beth Moreau, "Tahl Tan Mission Study," Opportunities for Youth Project, 1971.

MS2185, Ball Family Fonds, 1920–81.

MS2784, Tommy Walker Fonds, 1904–89.

Library and Archives Canada, Ottawa

RG10, Department of Indian Affairs, Black Series Collection.

Northern British Columbia Archives, Prince George

2000.1 Cassiar Asbestos Corporation Limited Fonds, 1952–93.

Prince Rupert Regional Archives, Prince Rupert

985–32 MS672, Dome Petroleum Files.

University of British Columbia Rare Book and Special Collections, Vancouver

RBSC-ARC-1069 Royal British Columbia Railway Commission Research Collection.

Yukon Archives, Whitehorse

81/105 Cassiar Asbestos Corporation Limited Fonds, 1951–78.

Case Law

Haida Nation v. British Columbia (Minister of Forests), [2004] 3 S.C.R. 511, 2004 SCC 73.

Taku River Tlingit First Nation v. British Columbia (Project Assessment Director), [2004] 3 S.C.R. 550, 2004 SCC 74.

Tsilhqot'in Nation v. British Columbia, [2014] SCC 44.

Secondary Sources

Abel, Kerry, and Ken Coates. *Northern Visions: New Perspectives on the North in Canadian History*. Peterborough, ON: Broadview Press, 2001.

Affleck, Edward R. "Navigation on the Stikine River, 1862–1969." *British Columbia Historical News* 28, 1 (1994–95): 15–19.

Ahlbom, Anders, and Maria Feychting. "Electromagnetic Radiation: Environmental Pollution and Health." *British Medical Bulletin* 68, 1 (2008): 157–65.

Alaska–Canada Energy Coalition. http://acecoalition.com/index.html.

Alaska Canada Rail Link. *Alaska Canada Rail Link Newsletter* 1, January 2006.

Alaska Canada Rail Link. "Alternate Route Segment Assessment, Yukon and BC, Final Report," June 2006.

"Alaskan Road Plan Aids BC: $354 Million Project." *Province,* April 1, 1961.

"Alaskans Fear Loss of Rich Estuary Habitat." *Telkwa Foundation Newsletter* 4, 2 (1981): 3.

Albright, Sylvia. *Tahltan Ethnoarchaeology*. Burnaby, BC: SFU Department of Anthropology Publication no. 15, 1984.

Alcan RailLink. "Rails to Resources to Ports: The Alaska Canada Rail Link Project, Phase 1 Feasibility Study, Executive Report." Whitehorse, March 2007.

Altagas. http://www.altagas.ca/.

Andrews, Clarence. *Wrangell and the Gold of the Cassiar.* Seattle: Luke Tinker, 1937.

Annett, William. "Shifting Fortunes," *BC Business* 17, 2 (1989): 26–33.

Aresco. *Preliminary Archaeological Study of the Proposed Stikine-Iskut Hydroelectric Development, Draft Report.* Vancouver, April 1980.

Aspect Consultants and the Association of United Tahltans. *Stikine Basin Resource Analysis: An Evaluation of the Impacts of Proposed Hydroelectric Developments, Vol. 1.* Vancouver, 1981.

Baker, Douglas C. "Mining in Northern British Columbia." *Western Geography* 12 (2002): 1–12.

Baldwin, Andrew, Laura Cameron, and Audrey Kobayashi, eds. *Rethinking the Great White North: Race, Nature, and the Historical Geographies of Whiteness in Canada.* Vancouver: UBC Press, 2011.

Ball, Georgiana. "Stikine History: Freighting to Watson Lake Aerodrome along the Stikine-Dease Corridor, 1941–43." Telegraph Creek, BC: Stikine Community Association, 1992.

Barnes, Trevor. "Borderline Communities: Canadian Single Industry Towns, Staples, and Harold Innis." In *B/ordering Space*, edited by Henk Van Houtum, O. Kramsch, and W. Zierhofer, 109–22. Burlington, UK: Ashgate Publishing, 2005.

Barnes, Trevor, Roger Hayter, and Elizabeth Hay. "Stormy Weather: Cyclones, Harold Innis and Port Alberni." *Environment and Planning A* 33, 12 (2001): 2127–47.

Bassett, Lamont. "A River for the Taming: Megaplans in the Great Stikine Wilderness." *Harrowsmith* 8, 5 (1984): 32–45.

–. "Spatsizi: One of BC's Last, Great Wilderness Areas is at the Centre of an Ecopolitical Debate." *Western Living* 15, 7 (1985): 54–55, 57–58, 86–87.

Basso, Keith. *Wisdom Sits in Places: Landscape and Language among the Western Apache*. Albuquerque: University of New Mexico Press, 1996.

Bavington, Dean. *Managed Annihilation: An Unnatural History of the Newfoundland Cod Collapse*. Vancouver: UBC Press, 2010.

BC Hydro. *Hydroelectric Development of the Stikine River Overview Study*. Victoria: BC Hydro, Hydroelectric Design Division, 1978.

–. *Prospectus: Northern Transmission Studies; Electrical Energy Transmission from Potential Hydroelectric Generation Projects on the Stikine/Iskut and Liard Rivers*. Vancouver: BC Hydro, 1980.

–. *Stikine-Iskut Hydroelectric Development: Progress Report on Feasibility Studies*. Vancouver: BC Hydro, 1980.

–. *Stikine-Iskut Rivers Hydroelectric Project Environmental Feasibility Studies; Economic Geology, Generation Planning Department – System Engineering Department*. Vancouver: BC Hydro, 1980.

–. *Stikine-Iskut Hydroelectric Development Exploration Program and Access Requirements*. Vancouver: BC Hydro, 1981.

–. *Stikine-Iskut Hydroelectric Development Feasibility Study: Hydrology, River Regime and Morphology*. Vancouver: BC Hydro, 1981.

–. "Consultants Study Interplay of Stikine Estuary and River." *Northern Projects Journal* 1, 2 (1982).

–. *Preliminary Planning Report for the Stikine-Iskut Transmission System: A Summary of System Planning, Route Engineering, and Environmental Studies*. Vancouver: BC Hydro, 1982.

–. "Program Designed to Care for Environment." *Northern Projects Journal* 1, 2 (1982): 5, supplement to the *Prince George Citizen,* May 28, 1982.

–. *Stikine-Iskut Hydroelectric Development: Investigations Outline, 1983*. Vancouver: BC Hydro, Community Relations Department, 1983.

–. "Northwest Transmission Line Agreement Will Create Jobs, Power BC's Northwest." BC Hydro press release, May 28, 2010. https://www.bchydro.com/news/press_centre/news_releases/2010/ntl_agreement_will_create_jobs_power_bc_nw.html.

–. "Tahltan Nation Signs Agreement in Support of Northwest Transmission Line." BC Hydro press release, May 17, 2011. https://www.bchydro.com/news/press_centre/news_releases/2011/Thaltan_NTL_agreement.html.

BC Rail. "Dease Lake Rail Extension Study." In *Northern Extension: Fort St. James, Dease Lake and Fort Nelson: A Collection of Studies, 1961–1969*. Vancouver: British Columbia Railway, 1961–1969.

"BC's Mining Sector Pleased Northwest Transmission Line Approved." *Northern BC Business,* February 25, 2011. http://northernbcbusiness.com/2011/02/27/bcs-mining-sector-pleased-northwest-transmission-line-approved/.

BC Utilities Commission Hearings. *In the Matter of the* Utilities Commission Act *and in the Matter of an Application by BC Hydro and Power Authority Proceeding no. 48.* Vancouver: BCTC, June 1, 1982.

Beak Consultants. *Preliminary Analysis of the Potential Impact of Hydroelectric Development of the Stikine River System on Biological Resources of the Stikine Estuary, Prepared for BC Hydro.* Vancouver: Beak Consultants and BC Hydro, 1981.

Beanlands, Gordon E., and Peter N. Duinker. *An Ecological Framework for Environmental Impact Assessment in Canada.* Halifax, NS, and Hull, QC: Institute for Resource and Environmental Studies, Dalhousie University, and Federal Environmental Assessment Review Office, 1983.

–. "An Ecological Framework for Environmental Impact Assessment." *Journal of Environmental Management* 18, 3 (1984): 267–77.

Bell, W.H. "The Stikine River and Its Glaciers." *Scribner's Monthly* 27 (1879): 805–15.

Bennett, Jane. *Vibrant Matter: A Political Ecology of Things.* Chapel Hill, NC: Duke University Press, 2009.

Bennett, Nelson. "Northern Transmission Line Approved a $404 Million Transmission Line That Northern BC Communities Say Is Critical to Job Creation and Economic Development Has Been Given the Green Light." *Business in Vancouver,* May 9, 2011. http://www.bivinteractive.com/index.php?option=com_content&view=article&id=4187:northern-transmission-line-approved&catid=14:daily-news&Itemid=46.

Berger, Thomas R. *Northern Frontier, Northern Homeland: The Report of the Mackenzie Valley Pipeline Inquiry.* Ottawa: Minister of Supply and Services, 1977.

Berland, Jody. *North of Empire: Essays on the Cultural Technologies of Space.* Chapel Hill, NC: Duke University Press, 2009.

Biehler, Dawn Day. "Permeable Homes: A Historical Political Ecology of Insects and Pesticides in US Public Housing." *Geoforum* 40, 6 (2009): 1014–23.

Black, Samuel. *A Journal of a Voyage from Rocky Mountain Portage in Peace River to the Sources of Finlays Branch and North West Ward in Summer 1824.* Edited by E.E. Rich and A.M. Johnson. London: HBRS, 1955.

Blaikie, Piers M., and Harold Brookfield, eds. *Land Degradation and Society.* London and New York: Methuen, 1987.

Bocking, Stephen. *Nature's Experts: Science, Politics, and the Environment.* New Brunswick, NJ: Rutgers University Press, 2004.

–. "Science and Spaces in the Northern Environment." *Environmental History* 12, 4 (2007): 867–94.

Boyd, Robert. *The Coming Spirit of Pestilence: Introduced Infectious Diseases and Populations Decline among Northwest Coast Indians, 1774–1874.* Vancouver: UBC Press, 2000.

Bradbury, John. "Towards an Alternative Theory of Resource-Based Town Development in Canada." *Economic Geography* 55, 2 (1979): 147–66.

Bradbury, John, and Michael Sendbuehler. "Restructuring Asbestos Mining in Western Canada." *Canadian Geographer* 32, 4 (1988): 296–306.

Bradley, J.R. *Hunting Big Game in Far Northwestern British Columbia.* New York: Mail and Express Job Print, 1904.

Brannstrom, Christian. "What Kind of History for What Kind of Political Ecology?" *Historical Geography* 32 (2004): 71–87.

Braun, Bruce. "Producing Vertical Territory: Geology and Governmentality in Late-Victorian Canada." *Ecumene* 7, 1 (2000): 7–46.

–. *The Intemperate Rainforest: Nature, Culture and Power on Canada's West Coast.* Minneapolis: University of Minnesota Press, 2002.

Breen, David. *Alberta's Petroleum Industry and the Conservation Board.* Edmonton: University of Alberta Press, 1993.

Brian, Arthur W. *Increasing Returns and Path Dependence in the Economy.* Ann Arbor: University of Michigan Press, 1994.

Bridge, Gavin. "Contested Terrain: Mining and the Environment." *Annual Review of Environment and Resources* 29 (2004): 205–59.

–. "Gas, and How to Get It." *Geoforum* 35, 4 (2004): 395–97.

–. "Material Worlds: Natural Resources, Resource Geography and the Material Economy." *Geography Compass* 3, 3 (2009): 1217–44.

–. "The Territorialities of Liquefied Natural Gas." Paper presented at the 2013 AAG Annual Meeting, Los Angeles, California.

Bridge, Gavin, and Karen Bakker. "Material Worlds? Resource Geographies and the 'Matter of Nature.'" *Progress in Human Geography* 30, 1 (2006): 5–27.

Brinco Mining. *Cassiar Annual Report 1981.* Vancouver: Brinco Mining, 1981.

Brinco Mining and Wright Engineers. *Pre-Feasibility Study, Volume 1: Technical and Cost Report.* Vancouver: Brinco Mining, 1983.

British Columbia. "Factsheet: LNG Project Proposals in British Columbia", BC Gov. News, March 7, 2016, https://news.gov.bc.ca/factsheets/factsheet-lng-project-proposals-in-british-columbia.

British Columbia. "Agreement Reached on Land to Build LNG Facility at Grassy Point." BC Government Newsroom, November 12, 2013. https://news.gov.bc.ca/stories/agreement-reached-on-land-to-build-lng-facility-at-grassy-point.

British Columbia. Ministry of Energy and Mines. *British Columbia's Natural Gas Strategy: Fuelling BC's Economy for the Next Decade and Beyond.* Victoria: Ministry of Energy and Mines, 2013. http://www.gov.bc.ca/ener/popt/down/natural_gas_strategy.pdf.

British Columbia. Department of Lands. *Abstracts from Reports on the Peace River and Cassiar Districts, Made by British Columbia Land Surveyors to the Department of Lands, 1891–1928.* Victoria: King's Printer, 1929.

British Columbia. "LNG in BC – Aurora LNG." http://engage.gov.bc.ca/lnginbc/lng-projects/aurora-lng/.

British Columbia. Ministry of the Environment, "Northwest Transmission Line Approved," BC Gov. Information Bulletin, February 23, 2011. https://www.bchydro.com/content/dam/hydro/medialib/internet/documents/projects/ntl/NTL_info_bceao_approved_feb2011.pdf.

British Columbia. Ministry of Environment, Land and Parks, BC Parks Division. *Stikine Country Protected Areas Technical Background Information Summary, Draft Copy.* Victoria: BC Parks Division, 2000.

British Columbia. Ministry of Sustainable Resource Management. *Cassiar Iskut-Stikine Land and Resource Management Plan.* Victoria: Ministry of Sustainable Resource Management, 2000. https://www.for.gov.bc.ca/tasb/slrp/lrmp/smithers/cassiar/plan/files/CIS-LRMP-November-2006.pdf.

British Columbia Environmental Assessment Office. *2009 Aboriginal (A9) Comments on the NTL Project Draft Terms of Reference.* http://a100.gov.bc.ca/appsdata/epic/documents/

p299/1260404358142_5afbeac073c07828125ceed6a71cf5e955b8b3bf9ffc43dbd2f6cf4b9dd327df.pdf.

British Columbia Environmental Assessment Office. *2009 Public (P9) Comments on the NTL Project Draft Terms of Reference.* http://a100.gov.bc.ca/appsdata/epic/documents/p299/1260404496735_5afbeac073c07828125ceed6a71cf5e955b8b3bf9ffc43dbd2f6cf4b9dd327df.pdf.

British Columbia Transmission Corporation. *Northwest Transmission Line Project, Application for an Environmental Assessment Certificate Volumes I to IX.* Vancouver: Rescan Environmental Services, 2010.

–. *Northwest Transmission Line Project, Application for an Environmental Assessment Certificate, Part 1.* Vancouver: Rescan Environmental Services (Project #832–2), 2010.

Brody, Hugh. "Introduction." In *Stikine: The Great River,* ed. Gary Fiegehan. Vancouver: Douglas and McIntyre, 1991.

Brooks, Karl Boyd. *Public Power, Private Dams: The Hells Canyon High Dam Controversy.* Seattle: University of Washington Press, 2009.

Brooks, Paul. "The Plot to Drown Alaska." *Harper's Magazine* 215 (1965): 53–59.

Bryant, Raymond. "Political Ecology: An Emerging Research Agenda in Third-World Studies." *Political Geography* 11, 1 (1992): 12–36.

Bryant, Raymond, and Sinead Bailey. *Third World Political Ecology: An Introduction.* New York: Routledge, 1997.

Buckham, A. F., and B.A. Latour, The Groundhog Coalfield, British Columbia, Bulletin 16 of Geological Survey of Canada. E. Cloutier: King's Printer, 1950.

Buell, Frederick. "A Short History of Oil Cultures: Or, the Marriage of Catastrophe and Exuberance." *Journal of American Studies* 46, 2 (2012): 273–93.

Burgoyne, A.A. "Geology and Exploration, McDame Asbestos Deposit, Cassiar, BC." *The Canadian Mining and Metallurgical Bulletin* 79.889 (1986): 31–37.

Bustard, David R. *Environmental Problems Associated with the Abandonment of the Dease Lake Extension of BC Rail.* Smithers: BC Fish and Wildlife Branch, 1977.

Callison, Candis L. "A Digital Assemblage: Diagramming the Social Realities of the Stikine River Watershed." Master's thesis, MIT, 2002.

Cameron, Emilie. *Far Off Metal River: Inuit Lands, Settler Stories, and the Making of the Contemporary Arctic.* Vancouver: UBC Press, 2015.

Campbell, Claire. *Shaped by the West Wind: Nature and History in Georgian Bay.* Vancouver: UBC Press, 2007.

Campbell, Robert. *Two Journals of Robert Campbell.* Seattle: John W. Todd, 1958.

Campbell-Johnston, R.C. "The Importance of Groundhog Coal," *British Columbia Magazine* 9, 10 (October 1913): 571–76.

Canada. British Columbia-Yukon-Alaska Highway Commission. "Report on Proposed Highway through British Columbia and the Yukon Territory to Alaska." Ottawa: King's Printer, 1942.

Canada. National Energy Board. *Report of the Joint Review Panel for the Enbridge Northern Gateway Project* (December 19, 2013). http://gatewaypanel.review-examen.gc.ca/clf-nsi/dcmnt/rcmndtnsrprt/rcmndtnsrprt-eng.html.

Canada. "Official Report of the Debates of the House of Commons of the Dominion of Canada: Third Session, Eighth Parliament ... Comprising the Period from the Third Day of February to the Twenty-First Day of April Inclusive." Ottawa: S.E. Dawson, 1898.

Canadian Environmental Assessment Agency. "British Columbia Northwest Transmission Line Project: Decision." http://www.ceaa-acee.gc.ca/050/details-eng.cfm?evaluation=51726
Canadian Resourcecon and Inter-Island Coracle Consultation. *Social and Economic Impacts of the Western LNG Project.* Calgary: Dome Petroleum, 1982.
Careless, J.M.S. *Canada: A Story of Challenge.* Toronto: Macmillan, 1953.
Carlson, Hans. *Home Is the Hunter: The James Bay Cree and Their Land.* Vancouver: UBC Press, 2008.
Carter Energy. *Transpac Gas Project: Proposal, Liquefied Natural Gas (LNG) Export Project.* Vancouver: Carter Energy, 1981.
"Cassiar ... Do You Remember?" http://www.cassiar.ca/.
Cassiar Asbestos Company. *Annual Report 1952*. Toronto: Conwest Exploration Company, 1953.
–. *Annual Report 1953*. Toronto: Conwest Exploration Company, 1954.
–. *Annual Report 1954*. Toronto: Conwest Exploration Company, 1955.
–. *Annual Report 1957*. Toronto: Conwest Exploration Company, 1958.
–. *Annual Report 1961*. Toronto: Conwest Exploration Company, 1962.
–. *A Submission to the Pollution Control Board of British Columbia for the Public Inquiry.* Vancouver: Cassiar Asbestos, 1978.
"Cassiar Completes First Shipment." *Northern Miner,* February 28–March 5, 2000, 2.
Cassiar Courier 1, 1 (1976), 1–2.
Cassiar Mining Corporation. *Submission of Cassiar Mining Corporation Regarding Proposed Occupational Safety and Health Regulation of Asbestos.* Prepared for the Workers' Compensation Board of British Columbia. Vancouver: Cassiar Mining Corporation, 1989.
"Cassiar Output Starts in Fall." *Northern Miner,* July 31, 1952.
"Cassiar Road Agreement Reached." *Whitehorse Star,* November 20, 1958.
Cassidy, Frank, and Norman Dale. *After Native Claims? The Implications of Comprehensive Claims Settlements for Natural Resources in BC.* Montreal, QC: Institute for Public Policy, 1988.
Castle, William F. "Up the Stikine to the Cassiar." *Canadian Geographic Journal* 10, 1 (1935): 14–21.
Chabon, Michael. *Maps and Legends*. New York: McSweeney's, 2008.
Chastko, Paul. *Developing Alberta's Oil Sands: From Karl Clark to Kyoto*. Calgary: University of Calgary Press, 2004.
"Chretien Feels Japan Reassured about LNG." *Globe and Mail,* April 19, 1983.
Coates, Ken. *North to Alaska*. Anchorage: University of Alaska Press, 1992.
Coates, Ken, and William Morrison. *Land of the Midnight Sun: A History of the Yukon.* Edmonton: Hurtig Publishers, 1988.
Cole, Douglas. *Captured Heritage: The Scramble for Northwest Coast Artifacts.* Vancouver: UBC Press, 1995.
Connelly, Joel. "The Last Frontier: BC Considers Fate of Its Wild North." *Seattle Post-Intelligencer,* July 8, 1996: A1.
Conwest Exploration Company. *Annual Report 1951*. Toronto: Conwest Exploration Company, 1952.
"Corporate Donors Ask BC Liberals for Half-Billion-Dollar Handout." *Dogwood Initiative Bulletin,* February 20, 2007.
Cox, Kevin. "Consortium Try to Salvage LNG Deal with Japanese Firms." *Globe and Mail,* March 19, 1985, B3.

Crawford, Mark. *The Role of the State in the Economic Development of British Columbia: Case Studies of the Dease Lake Extension and Sukunka Coal.* Victoria: BC Project, University of Victoria (Political Science), 1983.

Creighton, Donald. *The Commercial Empire of the St. Lawrence* 1760–1850. Toronto: Ryerson Press, 1937.

Cronon, William. "Kennecott Journey: The Paths Out of Town." In *Under an Open Sky: Rethinking America's Western Past,* edited by William Cronon, G. Miles, and J. Gitlin, 28–51. New York: W.W. Norton, 1992.

Cruikshank, Julie. *Reading Voices: Oral and Written Interpretations of Yukon's Past.* Vancouver and Toronto: Douglas and McIntyre, 1991.

–. *Life Lived Like a Story: Life Stories of Three Yukon Elders.* Vancouver: UBC Press, 1998.

–. *The Social Life of Stories: Narrative and Knowledge in Northern Canada.* Vancouver: UBC Press, 1998.

–. *Do Glaciers Listen? Local Knowledge, Colonial Encounters, and Social Imagination.* Vancouver: UBC Press, 2005.

Curtis, Bruce. *The Politics of Population: State Formation, Statistics and the Census of Canada, 1840–1875.* Toronto: University of Toronto Press, 2001.

David, Paul A. *Technical Choice, Innovation and Economic Growth: Essays on American and British Experience in the Nineteenth Century.* Cambridge: Cambridge University Press, 1975.

Davis, Diana K. "Historical Approaches to Political Ecology." In *Handbook of Political Ecology,* edited by Gavin Bridge, J. McCarthy, and T. Perrault, 263–75. London: Routledge, 2015.

Davis, Wade. "Deep North." *National Geographic* 205, 3 (2004): 102–21.

–. *The Sacred Headwaters: The Fight to Save the Stikine, Skeena, and Nass.* Vancouver: Greystone Books, 2011.

–. "A Power Line to Nowhere: What Coal Mining Means to British Columbia's Sacred Headwaters." *The Walrus,* December 18, 2013. http://thewalrus.ca/a-power-line-to-nowhere/.

Dawson, George Mercer. *Report on an Exploration Made in the Yukon District, N.W.T. and Adjacent Portion of British Columbia, 1887.* Ottawa: Geological Survey of Canada, 1898.

Day, Chad Norman. "Update from Chad Day about Mount Polley Mine." *Tahltan Central Council,* August 6, 2014. http://tahltan.ca/update-chad-day-mount-polley-mine/.

Demchuk, Andrea. "The Stikine: Tahltans, Environmentalists and BC Hydro." Master's thesis, University of British Columbia, 1985.

Demeritt, David. "Ecology, Objectivity, and Critique in Writings on Nature and Human Societies." *Journal of Historical Geography* 20, 1 (1994): 22–37.

Demerjian, Bonnie. *Roll On! Discovering the Wild Stikine River.* Wrangell, AK: Stikine River Books, 2006.

Desbiens, Caroline. *Power from the North: Territory, Identity, and the Culture of Hydro-electricity in Quebec.* Vancouver: UBC Press, 2013.

District of Kitimat. *Comparative Evaluation for Construction of a Liquefied Natural Gas Plant in Kitimat (Bish Cove), or Prince Rupert-Port Simpson (Grassy Point), BC* District of Kitimat, 1982.

Doelle, Meinhard, and A. John Sinclair. "Time for a New Approach to Public Participation in EA: Promoting Cooperation and Consensus for Sustainability." *Environmental Impact Assessment Review* 26, 2 (2006): 185–205.

Dome Petroleum. *Dome, LNG and British Columbia: Initial Perspectives on Dome Petroleum's Proposed Western Liquefided [sic] Natural Gas Project.* Calgary: Dome Petroleum, 1981.

–. "Western LNG Project Newsletter, Fifth Edition." Calgary: Dome Petroleum, 1982: 1–8.

–. *Volume Three: Environmental Setting and Assessment for a Liquefied Natural Gas Terminal Grassy Point, Port Simpson Bay Northern British Columbia – Revision April 30, 1983.* Calgary: Dome Petroleum, 1981, 1983.

"Dome's LNG Plan Criticized for Costs Exceeding Benefits." *Globe and Mail,* October 9, 1982, B16.

"Dome Might Leave Alsands." *New York Times,* February 5, 1982. http://www.nytimes.com/1982/02/05/business/dome-might-leave-alsands.html

Dome Petroleum and Lax Kw'alaams Indian Band Council. *Agreement.* Vancouver: Dome Petroleum, 1983.

"Dome Signs Shipyard Pact with Japanese." *Globe and Mail,* February 25, 1981, B15.

"Dome Talks on Asset Sale." *New York Times,* August 5, 1982. http://www.nytimes.com/1982/08/05/business/dome-talks-on-asset-sale.html.

Doran, Gerry. "Environmental – Cassiar." *Cassiar Reporter* 1 (1976): 5.

Doughty-Davies, J. *Power Possibilities of Nass and Stikine Rivers.* Victoria: Water Rights Branch, 1955.

Downs, Art. *Paddlewheels on the Frontier: The Story of British Columbia and Yukon Sternwheel Steamers.* Seattle: Superior Publishing, 1972.

Drolet, Jean-Paul. "The Demands and Limitations Governing the Mineral Production of Western Canada to 1990." Presented at the Plenary Session on the Extraction of Our Mineral Resources, Vancouver, October 13, 1976.

Duinker, Peter N., and Lorne A. Greig. "The Impotence of Cumulative Effects Assessment in Canada: Ailments and Ideas for Redeployment." *Environmental Management* 37, 2 (2006): 153–61.

Dupont, V.H. "Report of an Explanation of the Upper Part of the Stikine River to Ascertain the Feasibility of a Railway." Canada, Department of Railways and Canals, Annual Report for Fiscal Year July 1, 1899–June 30, 1900, Sessional Paper No. 20, 1901: 148–72.

Dusyk, Nichole. "Downstream Effects of a Hybrid Forum: The Case of the Site C Hydroelectric Dam in British Columbia, Canada." *Annals of the Association of American Geographers* 101, 4 (2011): 873–81.

Dyce, Matt. "Canada between the Photograph and the Map: Aerial Photography, Geographical Vision and the State." *Journal of Historical Geography* 39, 1 (2013): 69–84.

Ecology and Environment. *LNG Risk Analysis – Western LNG Project.* Calgary: Dome Petroleum, 1981.

Elliott, Thomas A. "A Strategic Road to the North." *Western Business & Industry* (1964): 20–21.

Emmons, George T. *The Tahltan Indians.* University of Pennsylvania, the Museum, Anthropological Publication, vol. 4, , no.1, 1911.

"Employee Information Booklet." Cassiar Asbestos Corporation, N.D.

Enarson, D.A., V. Embree, L. MacLean and S. Gryzbowski. "Respiratory Health in Chrysotile Asbestos Miners in British Columbia: A Longitudinal Study." *British Journal of Industrial Medicine* 45, 7 (1988): 459–63.

"Energy Price Drop Will Not Halt LNG Project." *Globe and Mail,* March 25, 1983, B3.

Envirocon and Pearse-Bowden Economic Consultants. *The Socioeconomic Effects of the BC Rail's Dease-Lake Extension on the Stuart-Trembleur Lakes Indian Band*, Vancouver: Report for British Columbia Railway Company, February 1974.

Evenden, Matthew. *Fish Versus Power: An Environmental History of the Fraser River.* Cambridge: Cambridge University Press, 2004.

–, ed. "Site C Forum." *BC Studies* 161 (2009): 93–114.

Farish, Matt, and P. Whitney Lackenbauer. "High Modernism in the North: Planning Frobisher Bay and Inuvik." *Journal of Historical Geography* 35, 3 (2009): 517–44.

Farrow, Moira. "BC Plans Five Hydro Dams: Three New Towns." *Vancouver Sun*, December 17, 1973.

Faustmann, John. "The Future of the Stikine Basin." *Special Affairs Issue, BC Region* 1, 1, Vancouver, 1982.

–. "The Endangered Stikine." *Western Living* 12, 2 (1982): 32–36.

Feit, Harvey A. *Protecting Indigenous Hunters: The Social and Environmental Protection Regime in the James Bay and Northern Quebec Land Claims Agreement.* University of Michigan, Natural Resources Sociology Research Lab, 1982.

Ferguson, Barry. *Athabasca Oil Sands: Northern Resource Exploration, 1875–1951.* Regina: Canadian Plains Research Center, 1986.

Ferguson, James. *The Anti-Politics Machine: "Development," Depoliticization, and Bureaucratic Power in Lesotho.* Cambridge: Cambridge University Press, 1990.

Fidler, Courtney Riley. "Aboriginal Participation in Mineral Development: Environmental Assessment and Impact and Benefit Agreements." Master's thesis, University of British Columbia, 2008.

Follett, Amanda. "Feds Cutting Corners on Environmental Assessments: Supreme Court." *The Tyee*, January 21, 2010. http://thetyee.ca/Blogs/TheHook/Environment/2010/01/21/RedMine/.

Forgues, E.L. "Western LNG Project." Paper presented to Canadian Institute of Energy, 1982.

Forsyth, Tim. *Critical Political Ecology: The Politics of Environmental Science.* London: Routledge, 2003.

Fortune Minerals. http://www.fortuneminerals.com/news/press-releases/press-release-details/2015/Fortune-Minerals-and-POSCAN-complete-sale-of-Arctos-coal-licenses-to-BC-Rail-with-a-10-year-repurchase-option/default.aspx

"Fortune Minerals and Tahltan Nation Enter into Environmental Assessment Cooperation Agreement for the Mount Klappan Anthracite Coal Project, British Columbia." *Canada NewsWire*, February 9, 2009. https://business.highbeam.com/1758/article-1G1-193303982/fortune-minerals-tahltan-nation-enter-into-environmental.

Foster, B.R., and E.Y. Rahs. "A Study of Canyon-Dwelling Mountain Goats in Relation to Proposed Hydroelectric Development in Northwestern British Columbia, Canada." *Biological Conservation* 33, 3 (1985): 209–28.

Foster, B.R., and E.Y. Rahs (Mar-Terr Enviro Research). *Relationship between Mountain Goat Ecology and Proposed Hydroelectric Development on the Stikine River, BC: Final Report on 1979–80 Field Studies, Prepared for BC Hydro (Generation Planning Department, System Engineering Division).* Vancouver: BC Hydro, 1981.

"$400 Monthly Earned in Northern Post." *Ottawa Citizen*, December 31, 1952.

Friends of the Stikine. *Newsletter* 1 (1981).

–. *Newsletter* 12 (1985).

Friesen, David E. "Aboriginal Settlement Patterns in the Upper Stikine River Drainage, Northwestern British Columbia." Master's thesis, University of Calgary, 1985.

Froschauer, Karl. *White Gold: Hydroelectric Power in Canada*. Vancouver: UBC Press, 1999.

Gaglardi, P.A. "Brief on Transportation by British Columbia." Paper presented at Alaska-Yukon-British Columbia Conference, Victoria, July 20, 1960.

Garland, Hamlin. *Trail of the Goldseekers*. New York: MacMillan, 1899.

Garvie, Kathryn H., and Karena Shaw. "Oil and Gas Consultation and Shale Gas in British Columbia." *BC Studies* 184 (2014–15): 73–102.

–. "Shale Gas Development and Community Response: Perspectives from Treaty 8 Territory, British Columbia." *Local Environment: The International Journal of Justice and Sustainability* (forthcoming). doi:10.1080/13549839.2015.1063043.

Genuis, Stephen J. "Fielding a Current Idea: Exploring the Public Health Impact of Electromagnetic Radiation." *Public Health* 122, 2 (2008): 113–24.

Gibson, Robert. "From Wreck Cove to Voisey Bay: The Evolution of Federal Environmental Assessment in Canada." *Impact Assessment and Project Appraisal* 20, 3 (2002): 151–59.

Gibson, Robert, Hugh Benevides, Meinhard Doelle, and Denis Kirchhoff. "Strengthening Strategic Environmental Assessment in Canada: An Evaluation of Three Basic Options." *Journal of Environmental Law and Practice* 20, 3 (2010): 175–211.

Gill, Robert, Greg Kula, Greg Wolfman, Jay Melnyk, and Dana Rogers. "Galore Creek Project, British Columbia: NI 43-101 Technical Report on Pre-Feasibility Study." Vancouver: NovaGold Resources and AMEC Americas, 2011. http://www.novagold.com/_resources/projects/technical_report_galore_creek.pdf.

Gillespie, Greg. *Hunting for Empire: Narratives of Sport in Rupert's Land, 1840–1870*. Vancouver: UBC Press, 2007.

Golder Associates, "Northwest Transmission Line: Environmental Assessment Certificate Application Review First Nations Consultation Summary." Prepared for BC Hydro Aboriginal Relations and Negotiations (Report no. 08–1477–0016), Victoria, 2010.

Goldman, Mara J., Paul Nadasdy, and Matthew D. Turner. *Knowing Nature: Conversations at the Intersection of Political Ecology and Science Studies*. Chicago: University of Chicago Press, 2011.

Goldman, Michael. *Imperial Nature: The World Bank and Struggles for Social Justice in the Age of Globalization*. New Haven, CT: Yale University Press, 2005.

Golinski, Jan. *Making Natural Knowledge: Constructivism and the History of Science*. Chicago: University of Chicago Press, 1998.

"Government of Canada Annouces $130 Million Investment in Northwest Transmission Line," *Northwest Development Inititaive Trust*, September 16, 2009, http://www.northerndevelopment.bc.ca/news/government-of-canada-announces-130-million-investment-in-northwest-transmission-line/

Govier, George W., and British Columbia. *Commissioner Inquiry on British Columbia's Requirements, Supply and Surplus of Natural Gas and Natural Gas Liquids*. Victoria: Ministry of Energy, Mines and Petroleum Resources, 1982.

Grandin, Greg. *Fordlandia: The Rise and Fall of Henry Ford's Forgotten City*. New York: Metropolitan Books, 2009.

Grant Thornton LLP. *Employment Impact Review*. Prepared for the BC Ministry of Energy, Mines and Natural Gas, February 2013.

Greer, Allan, and Ian Radforth, eds. *Colonial Leviathan: State Formation in Mid-Nineteenth Century Canada*. Toronto: University of Toronto Press, 1992.

Grove, Richard. *Green Imperialism: Colonial Expansion, Tropical Island Edens and the Origins of Environmentalism, 1600–1860*. Cambridge: Cambridge University Press, 1995.

Gunton, Thomas. "Megaprojects and Regional Development: Pathologies in Project Planning." *Regional Studies* 37, 5 (2003): 505–19.

GW Solutions. *Coal Bed Methane and Salmon: Assessing the Risks*. Calgary: Pembina Institute, 2008.

Hamilton, Gordon. "Gitanyow Agree to Northern Transmission Line through Their Territory." *Vancouver Sun*, June 15, 2011. http://vancouversun.com/news/staff-blogs/gitanyow-agree-to-northern-transmission-line-through-their-territory.

Hayball, Gwen. *Warburton Pike: An Unassuming Gentleman*. Poole, UK: Gwen Hayball, 1994.

Hayward, Ian, and Associates. *Stikine/Iskut: Preliminary Environmental and Social Impact Assessment of the Stikine-Iskut Transmission System, Prepared for BC Hydro*. Vancouver: BC Hydro, 1982.

Hedlin Menzies and Associates. *The Canadian Northwest Transportation Study, Final Report*. Prepared for Government of Canada, Ministry of Transport, 1970.

Heynen, Nik, James McCarthy, Scott Prudham, and Paul Robbins, eds. *Neoliberal Environments: False Promises and Unnatural Consequences*. London and New York: Routledge, 2007.

Hick, William. "An Introduction to Northwestern British Columbia; (and) Major Underdeveloped Water Powers of Northern British Columbia, Geo. J. Smith." In *An Introduction to Northwestern British Columbia: Transactions of the Eight British Columbia Natural Resources Conference*. Victoria: n.p.,1955.

"Highway Power Line Receives Federal Approval." *Highway 37 Coalition*, May 6, 2011. http://www.mining.bc.ca/news/highway-37-power-line-receives-federal-approval.

Hindlip, Charles Alsopp. "Hunting in British Columbia." *Travel and Exploration* 1, 3 (1909): 177–87.

"History of the Stewart-Cassiar Highway." *Cassiar Courier* 4, 2 (December, 1979), 2, 19.

Hoagland, Edward. *Notes from the Century Before: A Journal From British Columbia*. New York: Random House, 1969.

–. *Early in the Season: A British Columbia Journal*. Vancouver: Douglas and McIntyre, 2008.

Hornig, J.F., ed. *Social and Environmental Impacts of the James Bay Hydroelectric Project*. Montreal and Kingston: McGill-Queen's University Press, 1999.

Hughes, J. David. *Drill, Baby, Drill: Can Unconventional Fuels Usher in a New Era of Energy Abundance?* Santa Rosa, CA: Post Carbon Institute, 2013.

–. "BC LNG: A Reality Check." *Watershed Sentinel*, January 17, 2014. http://watershedsentinel.ca/files/files/Hughes-BC-LNG-Jan2014.pdf.

Hume, Mark. "BC's Sacred Headwaters to Remain Protected from Drilling," *Globe and Mail*, December 18, 2012, http://www.theglobeandmail.com/news/british-columbia/bcs-sacred-headwaters-to-remain-protected-from-drilling/article6504385/.

Hume, Mark, and Wendy Stueck. "Elders Stage Month-Long Sit-In Waiting to Speak to Chief." *Globe and Mail*, February 26, 2005, A5.

–. "Does BC's New Power Line Fall Short of Being Green?" *Globe and Mail*, September 21, 2009. http://www.theglobeandmail.com/news/national/does-bcs-new-power-line-fall-short-of-being-green/article1295110/.

–. "BC Mine Falling Through Very Large Crack in System." *Globe and Mail*, February 1, 2010, A8.

Hunter, Justine. "Harper Pledges $130 Million for Northern BC Power Line." *Globe and Mail,* September 16, 2009. http://www.theglobeandmail.com/news/national/british-columbia/harper-pledges-130-million-for-northern-bc-power-line/article1290208/.

–. "LNG's Greenhouse-gas Impact Is Prompting Clark to Get Creative with Numbers." *Globe and Mail,* November 17, 2013. http://www.theglobeandmail.com/news/british-columbia/lngs-greenhouse-gas-impact-is-prompting-clark-to-get-creative-with-numbers/article15480455/.

Imperial Metals. http://www.imperialmetals.com/our-operations-and-projects/operations/red-chris-mine/overview.

Innis, Harold. *The Fur Trade in Canada: An Introduction to Canadian Economic History.* New Haven, CT: Yale University Press, 1930.

–. *Settlement and the Mining Frontier*. Toronto: MacMillan, 1936.

–. "The Political Implications of Unused Capacity in Frontier Economies." In *Staples, Markets, and Cultural Change: Selected Essays,* edited by Daniel Drache, 24–34. Montreal and Kingston: McGill-Queen's University Press, 1995.

International Energy Agency. *Golden Rules for the Golden Age of Gas: World Energy Outlook, Special Report on Unconventional Energy*. Paris: IEA, 2012.

Isaac, Thomas, Tony Knox, and Sarah Bird. "The Crown's Duty to Consult and Accommodate Aboriginal Peoples: The Supreme Court of Canada's Decisions in *Haida Nation v. BC* and *Weyerhaeuser and Taku River Tlingit First Nation v. BC*." *Aboriginal Law Group* (2005): 1–8.

Isard, Philip. "Northern Vision: Northern Development during the Diefenbaker Era." Master's thesis, Wilfred Laurier University, 2010.

Jaccard, Mark, and Brad Griffin. *Shale Gas and Climate Targets: Can They Be Reconciled?* Victoria: Pacific Institute for Climate Solutions, 2010.

Jackson, Ian C. "A Territory of Little Value, the Wind of Change on the Northwest Coast, 1861–1867." *The Beaver* 298 (1967): 40–45.

–. "The Stikine Territory Lease and Its Relevance to the Alaskan Purchase." *Pacific Historical Review* 36, 3 (1967): 289–306.

Jasanoff, Sheila. *Designs on Nature: Science and Democracy in Europe and the United States*. Princeton, NJ: Princeton University Press, 2005.

–. *The Fifth Branch: Science Advisors as Policy Makers*. Cambridge, MA: Cambridge University Press, 1990.

Jorgensen, Dolly, and Sverker Sorlin, eds. *Northscapes: History, Technology, and the Making of Northern Environments*. Vancouver: UBC Press, 2013.

Josephson, Paul R. *Indistrialized Nature: Brute Force Technology and Transformation of the Natural World.* Washington, DC: Island Press, 2002.

Keeling, Arn, and John Sandlos. "Environmental Justice Goes Underground? Historical Notes From Canada's Northern Mining Frontier." *Environmental Justice* 2, 3 (2009): 117–25.

Kerr, Forrest. "Dease Lake Area, Cassiar District, BC." In *1925 Summary Report of the Canadian Dept of Mines, Geological Survey, Pt. A.* (Ottawa, 1926).

–. "Lower Stikine and Western Iskut River Areas, BC." Canada. Department of Mines and Resources, Geological Survey Memoir No. 246. Ottawa, 1948.

Kessel, Clive. "Community Involvement in 'Mega-Project' Planning: A Case Study of the Relationship between the Lax Kw'alaams Indian Band and Dome Petroleum." Master's thesis, University of British Columbia, 1984.

Kirsch, Scott. *Proving Grounds: Project Plowshare and the Unrealized Dreams of Nuclear Earthmoving.* New Brunswick, NJ: Rutgers University Press, 2005.
Knight, Nancy, Peter Boothroyd, Margaret Eberle, June Kawaguchi, and Christiane Gagnon. *What We Know About the Socio-Economic Impacts of Canadian Megaprojects: An Annotated Bibliography of Post-Project Studies.* Vancouver: UBC, Centre for Human Settlements, 1994.
Krugman, Paul. "Increasing Returns and Economic Geography." *Journal of Political Economy* 99, 3 (1991): 483–99.
Lackenbauer, P. Whitney, and Matt Farish. "The Cold War on Canadian Soil: Militarizing a Northern Environment." *Environmental History* 12, 4 (2007): 921–50.
Landry, Richard. "Telegraph Trail." *Alaska* 35, 11, and 36, 1 (1969 and 1970).
Landucci, Janet M. "An Environmental Assessment of the Effects of Cassiar Asbestos Corporation on Clinton Creek, Yukon Territory." Department of Environment, Environmental Protection Service, Pacific Region (Regional Program Report no. 79–13), 1978.
Larsen, Bruce. "Sun Writer Walks In on Crew: Secret Stikine Work Uncovered." *Vancouver Sun,* August 7, 1971, 1 and 14.
Lawrence, Guy. "We Came to Get Rich – Damn Good Reason!" *Alaska Sportsman* (1943): 6–7, 22–28.
–. *Forty Years on the Yukon Telegraph.* Quesnel, BC: Caryall Books, 1965.
Le Baron, Bentley. *Stikine-Iskut Project: Social and Economic Impacts, Prepared for BC Hydro and Power Authority.* Vancouver: BC Hydro, 1982.
Leblanc, Suzanne. *Cassiar: A Jewel in the Wilderness.* Prince George, BC: Caitlin Press, 2003.
Lee, Norman. *Klondike Cattle Drive: The Journal of Norman Lee.* Vancouver: Mitchell Press, 1960.
Lewiston, Jennifer. "NEB Told Dome Canadian Content in LNG Plan Might Not Reach 50%." *Globe and Mail,* October 26, 1982, B1.
Li, Tania Murray. "Beyond 'the State' and Failed Schemes." *American Anthropologist* 107, 3 (2005): 383–94.
–. *The Will to Improve: Governmentality, Development and the Practice of Politics.* Chapel Hill, NC: Duke University Press, 2007.
Linton, Jamie. *What Is Water? A History of a Modern Abstraction.* Vancouver: UBC Press, 2010.
Lizee, Erik. "Rhetoric and Reality: Albertans and Their Oil Industry under Peter Lougheed" Master's thesis, University of Alberta, 2010.
Loken, Marty. *The Stikine River.* Anchorage: Alaska Geographic, 1979.
Loo, Tina. *States of Nature: Conserving Canada's Wildlife in the Twentieth Century.* Vancouver: UBC Press, 2006.
–. "Disturbing the Peace: Environmental Change and the Scales of Justice on a Northern River." *Environmental History* 12, 4 (2007): 895–919.
Loo, Tina, and Meg Stanley. "An Environmental History of Progress: Damming the Peace and Columbia Rivers." *Canadian Historical Review* 92, 3 (2011): 399–427.
MacBride, W.D. "From Montana to the Klondyke." *Caribou and Northwest Digest* 7 (April 1951) 8–9, 16a-19; (May 1951), 6–9, 19–28; (June 1951), 20–35.
MacLachlan, Bruce B. "Notes on Some Tahltan Oral Literature." *Anthropologica* 4 (1957): 1–9.
–. "Tahltan." In *Handbook of North American Indians, Vol. 6: Subarctic,* ed. June Helm, 458–68. Washington, DC: Smithsonian Institution Press, 1981.

Malkinson, W.H., and Henry Wakabayashi. *North East Coal Development: Planning and Implementation.* Canadian Institute of Mining and Metallurgy and BC Ministry of Industry and Small Business Development, 1982.

Malloch, G.S. *Groundhog Coal Field: Geological Survey of Canada A Series Maps: 106A* (Geological Survey of Canada Memoir 69, 1913).

Martin, Douglas. "Dome's Fight to Untangle a Vast Financial Web." *New York Times,* August 19, 1984.

Martin, Jason Grek. "Making Settler Space: George Dawson, the Geological Survey of Canada and the Colonization of the Canadian West in the Late 19th Century." PhD diss., Queen's University, 2009.

Martin, Thibault, and Steven M. Hoffman, eds. *Power Struggles: Hydro Development and First Nations in Manitoba and Quebec.* Winnipeg: University of Manitoba Press, 2011.

Matthews, Ralph, and Nathan Young. "Development on the Margin: Development Orthodoxy and the Success of Lax Kw'alaams, British Columbia." *Journal of Aboriginal Economic Development* 4, 2 (2005): 100–8.

Maurer, Florian. "Fledgling Fishery Threatened." *Telkwa Foundation Newsletter* 3, 2 (1980).

Maxwell, G. Brett. *A Brief Outline of the Economics and Northern Trucking Operations.* Economics Staff Group, Northern Economic Development Branch, Department of Indian Affairs and Northern Development, 1971.

McCart, P.J., D.W. Mayhood, M.L. Jones, and G.J. Glova. *Stikine-Iskut Fisheries Studies 1979, Prepared for BC Hydro.* Vancouver: BC Hydro, 1980.

McCarthy, James. "First World Political Ecology: Lessons from the Wise Use Movement." *Environment and Planning A* 34, 7 (2002): 1281–1302.

McClelland, R.H. "Selection of Western LNG." Edited by Hansard. British Columbia. Legislative Assembly 32nd Parl. 4th Sess 1982 Legislative Session (July 15, 1982).

–. *Project Selection: The Natural Gas Allocation Process, Statement by the Honourable R.H. McClelland, Minister of Energy, Mines and Petroleum Resources.* Victoria: Ministry of Energy Mines and Petroleum Resources, 1982.

McCreary, Tyler Allan. "Struggles of the Tahltan Nation." *The Canadian Dimension* 39, 6 (2005): 14.

McIllwraith, Thomas. *"We Are Still Didene": Stories of Hunting and History from Northern British Columbia.* Toronto: University of Toronto Press, 2012.

Mertha, Andrew C., and William R. Lowry. "Unbuilt Dams: Seminal Events and Policy Change in China, Australia, and the United States." *Comparative Politics* 39, 1 (2006): 1–20.

Millar, Susan, and Don Mitchell. "Spectacular Failure, Contested Success: The Project Chariot Bioenvironmental Programme." *Cultural Geographies* 5, 3 (1998): 287–302.

Miller, Bill. *Wires in the Wilderness: The Story of the Yukon Telegraph.* Victoria: Heritage House Publishing, 2004.

Mining Association of British Columbia. *Report on the Electrification of the Highway 37 Corridor: A Discussion on the Potential Benefits of a New Power Transmission Line to Northwest British Columbia*, September 2008. http://www.mining.bc.ca/sites/default/files/resources/final_mabcreport_electrification_of_highway_37_oct1.pdf.

Mitchell, Timothy. *Rule of Experts: Egypt, Techno-Politics, Modernity.* Berkeley: University of California Press, 2002.

Muir, John. *Travels in Alaska.* New York: Houghton Mifflin, 1917.

Mulvihill, Peter R., Douglas C. Baker, and William R. Morrison. "A Conceptual Framework for Environmental History in Canada's North." *Environmental History* 6, 4 (October 2001): 611–26.

Nadasdy, Paul. *Hunters and Bureaucrats: Power, Knowledge, and Aboriginal-State Relations in the Southwest Yukon.* Vancouver: UBC Press, 2004.

Nash, Linda. "The Changing Experience of Nature: Historical Experiences with a Northwest River." *The Journal of American History* 86 (2000): 1600–29.

Naske, Claus M. "The Taiya Project." *BC Studies* 91/92 (1991): 5–50.

Nay, C.S. "Stream Geochemical Reconnaissance in the Canadian Cordillera." Vancouver: Kennco Explorations (Western), n.d.

Nelles, H.V. *The Politics of Development: Forests, Mines and Hydro-Electric Power in Ontario, 1849–1941.* Montreal and Kingston: McGill-Queen's University Press, 1974.

Nexen CNOOC/Aurora LNG. http://auroralng.nexencnoocltd.com.

Noble, Bram. *Introduction to Environmental Impact Assessment: A Guide to Principles and Practice.* Don Mills, ON: Oxford University Press, 2006.

–. "Cumulative Environmental Effects and the Tyranny of Small Decisions: Towards Meaningful Cumulative Effects Assessment and Management." *Natural Resources and Environmental Studies Institute Occasional Paper* 8. University of Northern British Columbia, Prince George, BC, 2010.

North, Janine. "Northern BC Communities Electrified over Federal $130 Million Investment in the Northwest Transmission Line." *Northern Development Initiative Trust,* September 17, 2009. http://northerndevelopment.bc.ca/news/71/18/Northern-BC-Communities-Electrified-Over-Federal-130-Million-Investment-in-the-Northwest-Transmission-Line.

Northwest Power Line Coalition. *Highway 37 Report.* Terrace, BC, May 2009.

"Northwest Transmission Line Gets Environmental Green Light." *Northern View,* February 23, 2011. http://www.thenorthernview.com/news/116777064.html.

NovaGold Resources. http://novagold.com/properties/galore_creek/overview/?pageid=22238.

Nye, David. *Consuming Energy: A Social History of American Energies.* Cambridge, MA: MIT Press, 1998.

–. *Electrifying America: Social Meanings of a New Technology, 1880–1940.* Cambridge, MA: MIT Press, 1992.

O'Connor, Ryan. *The First Green Wave: Pollution Probe and the Origins of Environmental Activism in Ontario.* Vancouver: UBC Press, 2014.

O'Faircheallaigh, Ciaran. "Public Participation and Environmental Impact Assessment: Purposes, Implications, and Lessons for Public Policy Making." *Environmental Impact Assessment Review* 30, 1 (2010): 19–27.

Oberdeck, Kathryn J. "Archives of the Unbuilt Environment: Documents and Discourses of Imagined Space in Twentieth-Century Kohler, Wisconsin." In *Archive Stories: Facts, Fictions, and the Writing of History,* edited by Antoinette Burton, 251–74. Chapel Hill, NC: Duke University Press, 2006.

Offen, Karl. "Historical Political Ecology: An Introduction." *Historical Geography* 32 (2004): 7–18.

Office of the Prime Minister of Canada. "PM Announces Canada's Investment in Northwest Transmission Line." Office of the Prime Minister press release, September 16, 2009.

Ontario. *Report of the Royal Commission on Matters of Health and Safety Arising from the Use of Asbestos in Ontario*. Ontario: Ontario Ministry of the Attorney General, 1984.

Osborn, Stephen G., Avner Vengosh, Nathaniel R. Warner and Robert B. Jackson. "Methane Contamination of Drinking Water Accompanying Gas-Well Drilling and Hydraulic Fracturing." *PNAS* 108, 20 (2011): 1–5.

Oswalt, George. "Yukon Dust." *Transport Times* 2, 6 (1971).

Overstall, Richard. "Is There a Real Need for the Stikine-Iskut Dams?" *Telkwa Foundation Newsletter* 3, 2 (1980): 5.

Overton, James. "Uneven Regional Development in Canada: The Case of Newfoundland." *Review of Radical Political Economics* 10, 3 (1978): 106–16.

Page, Justin. *Tracking the Great Bear: How Environmentalists Recreated British Columbia's Coastal Rainforest*. Vancouver: UBC Press, 2014.

Pan Canadian Consultants. *Environmental Overview of the Dome Petroleum Limited Western LNG Pipeline, a Report Prepared for Westcoast Transmission Company Limited and Dome Petroleum Limited*. Calgary: Dome Petroleum, 1981.

Parker, Paul. "Canada-Japan Coal Trade: An Alternative Form of the Staple Production Model." *The Canadian Geographer* 41, 3 (1997): 248–67.

Parr, Joy. *Sensing Changes: Technologies, Environments and the Everyday 1953–2003*. Vancouver: UBC Press, 2009.

Parr, Joy, Jessica van Horssen, and John van der Veen. *Megaprojects*. University of Western Ontario, 2008. http://megaprojects.uwo.ca.

–. "The Practice of History Shared Across Differences: Needs, Technologies, and Ways of Knowing in the Megaprojects New Media Project." *Journal of Canadian Studies* 43, 1 (2009): 35–58.

Pasqualetti, Martin J. "Social Barriers to Renewable Energy Landscapes." *Geographical Review* 101, 2 (2011): 201–23.

–. "The Changing Energy Landscapes of North America." Energy and Environment Specialty Group (EESG) Plenary Lecture, 2013 AAG Annual Meeting, Los Angeles, California.

Patterson, David Stewart. "Dome Seeks Provincial Pacts to Beat Gas Export Deadline." *Globe and Mail*, February 2, 1983, B7.

Patterson, Raymond M. *Trail to the Interior*. Toronto: MacMillan of Canada, 1966.

Paulson, Monte. "A Gentle Revolution." *The Walrus* 2, 10 (2006): 64–75.

Paulson, Susan, Lisa L. Gezon, and Michael Watts. "Locating the Political in Political Ecology: An Introduction." *Human Organization* 62, 3 (2003): 205–17.

Peet, Richard, and Michael Watts. "Introduction: Development Theory and Environment in an Age of Market Triumphalism." *Economic Geography* 68, 3 (1993): 227–53.

Peet, Richard, and Michael Watts, eds. *Liberation Ecologies: Environment, Development, Social Movement*. London and New York: Routledge, 1996.

Peluso, Nancy Lee. "What's Nature Got To Do With It? A Situated Historical Perspective on Socio-Natural Commodities." *Development and Change* 43, 1 (2012): 79–104.

Petro-Canada Exploration, Westcoast Transportation, and Mitsui and Company. *Rim Gas Project: Proposal to the Government of British Columbia*. Calgary: Petro-Canada Exploration, 1982.

Pewsey, Brian. "Public Address." *Cassiar Reporter* 1 (October 1976), 1.

Peyton, Jonathan. "Imbricated Geographies of Conservation and Consumption in the Stikine Plateau." *Environment and History* 17, 4 (2011): 555–81.

–. "Moving through the Margins: The 'All-Canadian' Route to the Klondike and the Strange Experience of the Teslin Trail." In *Ice Blink: Navigating Northern Environmental History,* edited by Stephen Bocking and Brad Martin. Calgary: University of Calgary Press, 2016.
Pike, Warburton. *The Barren Grounds of Northern Canada*. London: MacMillan and Company, 1892.
–. *Through the Subarctic Forest: Down the Yukon by Canoe in 1887*. New York: E. Arnold, 1896.
Piper, Liza. *The Industrial Transformation of Subarctic Canada*. Vancouver: UBC Press, 2009.
–. "Knowing Nature through History." *History Compass* 11/12 (2013): 1139–49.
Plumb, W.N. *The Geology of the Cassiar Asbestos Deposit.* Cassiar: Cassiar Asbestos Corporation, 1968.
Plummer, J., and R. Plummer. "Stikine Residents Speak: Locals Oppose Flooding." *Telkwa Foundation Newsletter* 3, 2 (1980): 6.
Pojar, Rosamund. "Transmission Links: More Links Moved into Place." *Telkwa Foundation Newsletter* 3, 3 (1980).
–. "Wildlife Impact: Only Hydro Knows." *Telkwa Foundation Newsletter* 3, 2 (1980).
Pollon, Christopher. "Time to Get 'Wacky' Again: The Northwest Transmission Line." *The Tyee,* September 21, 2008. http://thetyee.ca/News/2009/09/21/northwest transmission/.
–. "Gordon Campbell's $400 Million Power Line Bet." *The Tyee,* November 13, 2008. http://thetyee.ca/News/2008/11/13/Hwy37/.
–. "Report from the Edge of BC's Copper Rush." *The Tyee,* January 13, 2011. http://thetyee.ca/News/2011/01/13/Stikine/.
–. "BC's Tahltan and the Road to Power." *The Tyee,* January 14, 2011. http://thetyee.ca/News/2011/01/14/TahltanRoadToPower/.
–. "Alaska Power and the Bleeding of the Northwest." *The Tyee,* February 18, 2011. http://thetyee.ca/News/2011/02/18/AlaskaPower/.
–. "Northwest Power Line Grows, So Does Controversy." *The Tyee,* July 18, 2011. http://thetyee.ca/News/2011/07/18/NorthwestTransmissionLine/.
–. "Reinvent Environmental Assessment in BC, Say Critics," *The Tyee,* Novermber 8, 2012. http://thetyee.ca/News/2012/11/08/Reinvent-Environmental-Assessment-in-BC/.
"Power Line Fate in Hands of Gov't." *Terrace Standard,* January 12, 2011. http://www.bclocalnews.com/error/?errorURL=http%3A%2F%2F.
Pratt, Larry. *The Tar Sands: Syncrude and the Politics of Oil.* Edmonton: Hurtig, 1976.
Pratt, Larry, and John Richards. *Prairie Capitalism: Power and Influence in the New West.* Toronto: McClelland and Stewart, 1979.
Pratt, Larry, and Ian Urquhart. *The Last Great Forest: Japanese Multinationals and Alberta's Northern Forests*. Edmonton: NeWest Press, 1994.
Princeton Mining Company. *Reorganization Plan – Summary*. Vancouver: Princeton Mining Company, 1991.
Prudham, W. Scott. "Poisoning the Well: Neo-liberalism and the Contamination of Municipal Water in Walkerton, Ontario." *Geoforum* 35, 3 (2004): 343–59.
Pynn, Larry. *The Forgotten Trail: One Man's Adventures on the Canadian Route to the Klondike.* Toronto: Doubleday Canada, 1996.
Raffles, Hugh. *In Amazonia: A Natural History*. Princeton, NJ: Princeton University Press, 2002.

Rajala, Richard. "'Streams Being Ruined from a Salmon Producing Standpoint': Clearcutting, Fish Habitat, and Forest Regulation in British Columbia, 1900–45." *BC Studies* 176 (2012–13): 93–132.

Read Environmental and Planning Associates. *Impact Study Grassy Point LNG Project, Vol. 1–5*. Report prepared for the Port Simpson Indian Band. Vancouver, 1982.

"Red Chris Mine: An Environmental Law Victory Can Still Be a Loss for the Environment." *West Coast Environmental Law Newsletter*, January 21, 2010. http://wcel.org/resources/environmental-law-alert/red-chris-mine-environmental-law-victory-can-still-be-loss-environ.

Reitz, C.R. *Brief to Be Presented at the Pearse Commission Hearings, July 24–25, 1981*. Appendix I in B. Le Baron, *Stikine-Iskut Project*.

Robb, Stewart Andrew. "The Collins Overland or Russian Extension Telegraph Project." Master's thesis, Simon Fraser University, 1966.

Robbins, Paul. *Political Ecology: A Critical Introduction*. London: Routledge, 2004.

Roosevelt, Theodore. *The Winning of the West*, vols. 1–4. New York: Putnam and Sons, 1889–1896.

Rose, Ramona. "Ghost Town Turns Virtual Town: Preserving Cassiar's History via Social Networking Sites." Paper presented at Canadian Historical Association Meeting, May 30, 2011, Fredericton, NB.

Rosenberg, Frantz. *Big Game Shooting in British Columbia*. London: Hopkinson and Company, 1928.

Ryan, William E. "BC Rail-to-Alaska Deal May Be Set Next Week: At Hush-Hush Conference." *Province*, August 29, 1959.

Sabin, Paul. "Voices from the Hydrocarbon Frontier: Canada's Mackenzie Valley Pipeline Inquiry (1974–1977)." *Environmental History Review* 19, 1 (1995): 17–48.

Sandlos, John. *Hunters at the Margins: Native People and Wildlife Conservation in the Northwest Territories*. Vancouver: UBC Press, 2007.

Sandlos, John, and Arn Keeling. "Claiming the New North: Mining and Colonialism at the Pine Point Mine, Northwest Territories, Canada." *Environment and History* 18, 1 (2012): 5–34.

–. "Zombie Mines and the (Over)burden of History." *Solutions Journal* 4, 3 (2013): 80–83.

Sandwell, Ruth, ed. *Powering Up Canada: The History of Power, Fuel and Energy from 1600*. Montreal, QC: McGill-Queen's University Press, 2016.

Schnurr, Matthew A. "Lowveld Cotton: A Political Ecology of Agriculture Failure in Natal and Zululand, 1884–1948." PhD diss., University of British Columbia, 2008.

Scott, James C. *Seeing Like a State: How Certain Schemes to Improve the Human Condition Have Failed*. New Haven, CT: Yale University Press, 1999.

"Shell's Suspension in Klappan Extended for Another Year." *Terrace Standard*, February 15, 2011. http://www.terracestandard.com/news/116244339.html.

Sigurdson, Albert. "British Columbia Gas Projects Could Mean New Industry in the Making." *Globe and Mail*, March 8, 1982, B14.

–. "LNG Price to Japan Declines." *Globe and Mail*, April 26, 1983, B2.

–. "Postponed LNG Delivery Called Disruptive to Buyers." *Globe and Mail*, October 21, 1983, B10.

Silvey, Rachel, and Katherine Rankin. "Development Geography: Critical Development Studies and Political Geographic Imaginaries." *Progress in Human Geography* 35, 5 (2011): 696–704.

Simpson, Scott. "Collapse of Galore Mine Project Leaves Transmission Line in Limbo." *Vancouver Sun,* November 28, 2007. http://www.canada.com/vancouversun/news/business/story.html?id=12ec5fd0-604c-4758-83d6-6859792efdf8.

–. "BC Power Line Plans Have Alaskans Buzzing." *Vancouver Sun,* C1, January 13, 2010.

Sinclair, A. John, and Meinhard Doelle. "Using Law as a Tool to Ensure Meaningful Public Participation in Environmental Assessment." *Journal of Environmental Law and Practice* 12, 1 (2003): 27–53.

Singer, Daniel J. *Big Game Fields of America, North and South.* London: Hodder and Stoughton, 1916.

Smith, Joseph Gordon. "British Columbia-Alaska Highway." *Pacific Travel Monthly* 1, 2 (1936): 10–15.

Smith, L.G. "Taming BC Hydro: Site C and the Implementation of the BC *Utilities Commission Act.*" *Environmental Management* 12, 4 (1990): 429–43.

Stephenson, Eleanor, Alexander Doukas, and Karena Shaw. "Greenwashing Gas: Might a 'Transition Fuel' Label Legitimize Carbon-Intensive Natural Gas Development?" *Energy Policy* 46 (2012): 452–59.

Stoler, Ann Laura. *Along the Archival Grain: Epistemic Anxieties and Colonial Common Sense.* New Haven, CT: Princeton University Press, 2010.

Stone, Andrew J. *Journals of Andrew J. Stone: Expeditions of Arctic and Subarctic America after Wild Sheep, Grizzly, Caribou, and Muskoxen.*" Edited by Margaret R. Frisina. Long Beach, CA: Safari Press, 2010.

"Study Encouraging for Cassiar." *Whitehorse Star,* April 24, 1987.

Stueck, Wendy. "New Power Transmission Line Could Spur Mining Development." *Globe and Mail,* October 2, 2007, S3.

Stunden Bower, Shannon. *Wet Prairie: People, Land and Water in Agricultural Manitoba.* Vancouver: UBC Press, 2011.

Suzuki, David. "Foreward." In *The Sacred Headwaters: The Fight to Save the Stikine, Skeena, and Nass,* by Wade Davis, vii–viii. Vancouver: Greystone Books, 2011.

Tafler, Sid. "Struggle Over Site of $4 Billion Plant Angers BC Towns." *Globe and Mail,* January 31, 1983, 10.

Tahltan Central Council, "Tahltan Nation Approves of Northwest Transmission Line Agreements in Historic Vote" Tahltan Central Council Press Release, April 18, 2011. http://thetyee.ca/News/2011/07/18/TAHLTAN-PR2_2011-Apr-18.pdf.

Tahltan First Nation and International Institute of Sustainable Development (IISD). *Out of Respect: The Tahltan, Mining, and the Seven Questions to Sustainability.* Report of the Tahltan Mining Symposium, April 4–6, 2003, Dease Lake, BC.

Tahltan Nation. *1910 Declaration of the Tahltan Tribe.* http://www.tndc.ca/pdfs/Tahltan%20Declaration.pdf.

Taylor, Lawrence D. "The Bennett Government's Pacific Northern Railway Project and the Development of British Columbia's 'Hinterland'" *BC Studies* 175 (2012): 35–56.

Taylor, Melvin S. "The Environmental Control Structure Required to Establish Corporate Policy, Communications, Corporate Environmental Management, and to Meet All Guidelines for a Safe and Healthy Environment at Cassiar Asbestos Corporation Limited in Cassiar BC." Prepared for Cassiar Asbestos Corporation, 1987.

Taylor, Paul. "Analysts Query Economics of Dome's LNG Scheme." *Globe and Mail,* February 1, 1983, B1.

–. "Dome Tries to Save Japanese Gas Deal." *Globe and Mail,* January 13, 1984.

Teit, James. "Notes on the Tahltan Indians of British Columbia." In *Boas Anniversary Volume: Anthropological Papers Written in Honor of Franz Boas on the Twenty-Fifth Anniversary of His Doctorate,* edited by Berthold Laufer, 337–49. New York: G.E. Stechert and Company, 1906.

–. "Two Tahltan Traditions." *Journal of American Folklore* 22 (1909): 314–18.

–. "On Tahltan (Athabaskan) Work, 1912." *Summary Reports of the Geological Survey of Canada,* Sessional Paper No. 26 (Ottawa, 1912): 484–87.

–. "Kaska Tales," *Journal of American Folklore* 30, 118 (1917): 427–73.

–. "Tahltan Tales." *Journal of American Folklore* 32, 124 (1919): 198–250.

–. "Tahltan Tales." *Journal of American Folklore* 34, 133 (1921): 223–53.

–. "Tahltan Tales." *Journal of American Folklore* 34, 134 (1921): 335–56.

–. "Field Notes on the Tahltan and Kaska Indians, 1912–1915." *Anthropologica* 3 (1956): 39–213.

Tennant, Paul. *Aboriginal Peoples and Politics: The Indian Land Question in British Columbia, 1849–1989.* Vancouver: UBC Press, 1990.

TERA Environmental Consultants. *Environmental Considerations in LNG Terminal Site Selection for Dome Petroleum Limited.* Calgary: Dome Petroleum, 1981.

"The Stewart-Cassiar Highway." *Cassiar Courier* (1979), 19.

Thomas, Shaun. "Aurora Signs an Exclusivity Deal for Grassy Point Development." *Northern View,* November 12, 2013. http://www.thenorthernview.com/news/231653041.html.

Thompson, Andrew R. British Columbia. *Statement of Proceedings: West Coast Oil Ports Inquiry.* Vancouver: West Coast Oil Ports Inquiry, 1978.

Thompson, Judy. *Recording Their Story: James Teit and the Tahltan.* Vancouver: Douglas and McIntyre, 2007.

"3.5-billion Liquefied-Natural-Gas Project Scrapped." *Montreal Gazette,* January 29, 1986, C3.

Trick, Bernice. "Town Reduced to Boxes of Documents." *Prince George Citizen,* June 9, 2000, 5.

Tripp, George A. "The Stikine Gateway and the Cassiar Region: A Study in Frontier Development." Master's thesis, University of Western Ontario, 1975.

Tsing, Anna. *Friction: An Ethnography of Global Connection.* Princeton, NJ: Princeton University Press, 2005.

Turkel, William. *The Archive of Place: Unearthing the Pasts of the Chilcotin Plateau.* Vancouver: UBC Press, 2007.

Turner, Frederick Jackson. *The Frontier in American History.* New York: Holt, 1921.

United States. *Message from the President of the United States Transmitting the Report of the Alaskan International Highway Commission,* Washington: Government Printing Office, 1940.

–. "Report of the Commission to Study the Proposed Alaska Highway to Alaska." Department of State, Conference Series No. 14. Washington: Government Printing Office, 1933.

Usher, Peter J. "Traditional Ecological Knowledge in Environmental Assessment and Management." *Arctic* 53, 2 (2000):183–93.

Utzig, G., M. Walmsley, and C. Clement (Pedalogy Consultants). *Biogeoclimatic Zonation of the Stikine Basin, Prepared for the Association of United Tahltans.* Calgary, November 1982.

van der Linden, J.R. "The Western LNG Project." *The Journal of Canadian Petroleum Technology* 22, 1 (1983): 55–60.

van Horssen, Jessica. *A Town Called Asbestos: Environmental Contamination, Health, and Resilience in a Resource Community*. Vancouver: UBC Press, 2016.

Van Nostrand, John. "If We Build It, They Will Stay." *The Walrus* (September 2014) http://thewalrus.ca/if-we-build-it-they-will-stay/.

Van Wyck, Peter. *The Highway of the Atom*. Montreal, QC: McGill-Queen's University Press, 2010.

Wedley, John R. "Laying the Golden Egg: The Coalition Government's Role in Post-War Northern Development." *BC Studies* 88 (1990–91): 58–92.

White, Richard. *Railroaded: The Transcontinental Making of Modern America*. New York: W.W. Norton, 2011.

Williams, A. Bryan. *Game Trails of British Columbia*. London: John Murray, 1925.

Willis, Roxanne. *Alaska's Place in the West: From the Last Frontier to the Last Great Wilderness*. Lawrence: University of Kansas Press, 2010.

Wilson, Jeremy. *Talk and Log: Wilderness Politics in British Columbia, 1965–96*. Vancouver: UBC Press, 1998.

Winson, Anthony. "The Uneven Development of Canadian Agriculture: Farming in the Maritimes and Ontario. *Canadian Journal of Sociology* 10, 4 (1985): 411–38.

Wong, Craig. "Miners Set to Benefit from New Power Line to Be Built in Northwestern BC" *Winnipeg Free Press,* May 15, 2011. http://www.winnipegfreepress.com/business/breakingnews/121852614.html.

Wright, Shelley. *Our Ice Is Vanishing/Sikuvut Nunguliqtuq: A History of Inuit, Newcomers, and Climate Change*. Montreal, QC: McGill-Queen's University Press, 2014.

Yergin, Daniel, and Michael Stoppard. "The Next Prize." *Foreign Affairs* 82, 6 (2003): 103–14.

Zalik, Anna. "Liquefied Natural Gas and Fossil Capitalism." *Monthly Review* 60, 6 (2008): 41–53.

Zaslow, Morris. *The Northward Expansion of Canada, 1914–1967*. Toronto: McClelland and Stewart, 1988.

Zelko, Frank. "Making Greenpeace: The Development of Direct Action Environmentalism in British Columbia." *BC Studies* 142/143 (2004): 241–77.

Zimmerman, Erich. *World Resources and Industries: A Functional Appraisal of the Availability of Agricultural and Industrial Materials*. New York: Harper, [1933] 1951.

Index

Note: "(f)" following a page number indicates a figure; "(t)" following a page number indicates a table. In subheadings, BCEAO refers to BC Environmental Assessment Office, LNG to liquified natural gas, THREAT to Tahltan Heritage Resources Environmental Assessment Team, and UNBC to the University of Northern British Columbia.

Claire Elizabeth Campbell, *Shaped by the West Wind: Nature and History in Georgian Bay*
Tina Loo, *States of Nature: Conserving Canada's Wildlife in the Twentieth Century*
Jamie Benidickson, *The Culture of Flushing: A Social and Legal History of Sewage*
William J. Turkel, *The Archive of Place: Unearthing the Pasts of the Chilcotin Plateau*
John Sandlos, *Hunters at the Margin: Native People and Wildlife Conservation in the Northwest Territories*
James Murton, *Creating a Modern Countryside: Liberalism and Land Resettlement in British Columbia*
Greg Gillespie, *Hunting for Empire: Narratives of Sport in Rupert's Land, 1840–70*
Stephen J. Pyne, *Awful Splendour: A Fire History of Canada*
Hans M. Carlson, *Home Is the Hunter: The James Bay Cree and Their Land*
Liza Piper, *The Industrial Transformation of Subarctic Canada*
Sharon Wall, *The Nurture of Nature: Childhood, Antimodernism, and Ontario Summer Camps, 1920–55*
Joy Parr, *Sensing Changes: Technologies, Environments, and the Everyday, 1953–2003*
Jamie Linton, *What Is Water? The History of a Modern Abstraction*
Dean Bavington, *Managed Annihilation: An Unnatural History of the Newfoundland Cod Collapse*
Shannon Stunden Bower, *Wet Prairie: People, Land, and Water in Agricultural Manitoba*
J. Keri Cronin, *Manufacturing National Park Nature: Photography, Ecology, and the Wilderness Industry of Jasper*
Jocelyn Thorpe, *Temagami's Tangled Wild: Race, Gender, and the Making of Canadian Nature*
Darcy Ingram, *Wildlife, Conservation, and Conflict in Quebec, 1840–1914*
Caroline Desbiens, *Power from the North: Territory, Identity, and the Culture of Hydroelectricity in Quebec*

Sean Kheraj, *Inventing Stanley Park: An Environmental History*
Justin Page, *Tracking the Great Bear: How Environmentalists Recreated British Columbia's Coastal Rainforest*
Daniel Macfarlane, *Negotiating a River: Canada, the US, and the Creation of the St. Lawrence Seaway*
Ryan O'Connor, *The First Green Wave: Pollution Probe and the Origins of Environmental Activism in Ontario*
John Thistle, *Resettling the Range: Animals, Ecologies, and Human Communities in British Columbia*
Carly A. Dokis, *Where the Rivers Meet: Pipelines, Participatory Resource Management, and Aboriginal-State Relations in the Northwest Territories*
Nancy B. Bouchier and Ken Cruikshank, *The People and the Bay: A Social and Environmental History of Hamilton Harbour*
Jessica Van Horssen, *A Town Called Asbestos: Environmental Contamination, Health, and Resilience in a Resource Community*

Printed and bound in Canada by Friesens
Set in Garamond by Artegraphica Design Co. Ltd.
Copy editor: Jillian Shoichet
Indexer: Marnie Lamb
Cartographer: Eric Leinberger
Cover designer: George Kirkpatrick